CW00591864

ISBN 978-1-332-73459-7
PIBN 10315566

ELEMENTARY MECHANICAL REFRIGERATION

A Simple and Non-Technical Treatise

BY

FRED. E. MATTHEWS, B. S., M E., E. E.

Member of American Society of Mechanical Engineers
and American Society of Refrigerating Engineers

FIRST EDITION
SECOND IMPRESSION

McGRAW-HILL BOOK COMPANY, INC.
239 WEST 39TH STREET, NEW YORK
6 BOUVERIE STREET, LONDON, E. C.

Printed by
The Maple Press
York, Pa.

PREFACE

The rapidly growing application of mechanical refrigeration, to new as well as to old industries, has given rise to the need of a more general understanding of its elementary principles. So rapid has been its development, in fact, that in the mechanical world both laymen, and specialists in other lines formerly having little or nothing in common with the art of refrigeration, have suddenly awakened to the fact that at least a casual understanding of the subject has become almost indispensable. While the theme has been a not unpopular one among writers of more or less technical text, yet an ample opportunity still remains to help the busy man at the desk, the drawing board and the throttle, to a better understanding of the principles involved and the methods employed. If the following exposition should be fortunate enough to save the valuable time of a busy man or lighten the burden of the overworked one, its purpose will not be unaccomplished.

In its brief life of less than one century the art of mechanical refrigeration has developed from the scattered and rather unscientific application of a few more or less imperfectly understood natural laws, to an industry of almost universal application and of importance second to none in the working out of the momentous problems of domestic, national, and international economy. To allow the manufacturer to produce better goods, to allow the farmer to market his products in better condition, to facilitate distribution, to lengthen the period of consumption and thereby eliminate market glut and famine; to provide more wholesome food for man—such is the beneficent work of mechanical refrigeration.

F. E. Matthews.

February, 1912.

CONTENTS.

CHAPTER V

CHAPTER VI

CHAPTER VII

CHAPTER VIII

CHAPTER IX

CHAPTER X

CHAPTER XI

ELEMENTARY MECHANICAL REFRIGERATION

CHAPTER I

COLD AND ITS PRODUCTION

Refrigeration the Extraction of Heat

In its broader sense refrigeration may be defined as the process of cooling; or, since cold is but the absence of heat, as darkness is absence of light, and dryness is absence of moisture, in which cases the real entities are heat, light, and moisture respectively, refrigeration may be more accurately defined as *the process of extracting heat.* The study of refrigeration therefore necessitates the study of heat.

The above definition of Refrigeration is somewhat inadequate, in that it conveys the impression that heat is a passive element, while in reality it is decidedly an active one. While heat may be generated by the performance of work, as in the compression of gases, or even when a piece of metal is struck a few sharp blows with a hammer (its appearance being incident to the disappearance of an equivalent amount of work), when once fortified within the walls of matter it is able to resist the most strenuous efforts to dislodge it and it must accordingly be decoyed into leaving the substance from choice.

Heat can best be coaxed out of a given substance by placing near it another substance materially lower in temperature, under which condition its tendency is to flow from the substance of higher to that of lower temperature—just as water flows from a higher to a lower level. The result of such a gravitation of heat is that the latter substance is heated and the former refrigerated. Wherever there is a difference in temperature between two bodies there is always a tendency for heat to flow from that of the higher to that of the lower temperature, hence it follows that both heating and refrigerating may take place at any point above absolute zero.

GENERAL THERMAL PROPERTIES OF MATTER

Since refrigeration has to do wholly with the extraction of heat, in order to arrive at a clear understanding of the subject one must first become familiar with the general thermal properties of the substances most commonly encountered in connection with refrigerating processes, especially such properties as have to do with their capacities for absorbing heat under different conditions.

THE MOLECULAR THEORY

The molecular theory of matter assumes that all the chemical elements, of which all matter is composed, are made up of infinitely small particles called *atoms.* When elements combine to form compounds, such as when hydrogen combines with oxygen to form water, the atoms of each element combine to form other infinitely small particles of matter called *molecules.* Two or more atoms may combine to form molecules. Atoms and molecules, although infinitely small, are supposed to possess all the properties of the respective substances of which they are a part. The application of a sufficient amount of heat to a substance will result in separating the molecules into their constituent atoms; or the substances (if a compound) into its elemental parts.

Matter may exist in three different states, the solid, the liquid, and the gaseous, according to the amount of heat that it contains.

THE KINETIC THEORY

The kinetic theory assumes that molecules and atoms are in constant motion. This motion is greatly restricted in solid bodies because of the powerful intermolecular attractive forces which exist when the molecules are so close together. The motion is less restricted in the liquid, and practically unrestricted in the gaseous state.

Heat is a form of energy. The application of heat to a substance increases the kinetic energy of its molecules, enabling them to increase their motion notwithstanding the strong attractive forces exerted to limit it. The addition of sufficient heat to a solid will so increase the energy of its molecules as to partially overcome the intermolecular attractive forces and allow of sufficiently increased motion to change the substance from the solid to the liquid state. A further application of heat energy will further overcome the attractive forces and change the substance from the liquid to the gaseous state.

In case of some complex substances, such for example as ammonia (NH_3) which is composed of the two elements, nitrogen (N) and hydrogen (H), a still further application of heat will effect the dissociation of its molecules into atoms of nitrogen and hydrogen.

In the general case the withdrawal of sufficient heat from a gas has the effect of causing it to return first to the liquid, then to the solid state.*

Change of Condition and State of Matter by Heat

Every substance absorbs a more or less constant quantity of heat for each degree rise in temperature in that substance. This quantity of heat expressed in proper units is known as the "specific heat of the solid," the "liquid" or the "gas," as the state of the substance in question may determine.

Each degree rise in temperature of a pound of a solid substance, for example, is attended by the absorption of a quantity of heat known as the "specific heat of the solid." The absorption of a practically fixed quantity of heat for each degree rise in temperature will continue until the melting point of the solid is reached—when a further application of heat will produce no further rise in temperature but will produce a change of state, liquefaction, instead. The amount of heat absorbed per pound during liquefaction is known as the "latent heat of fusion" or the "latent heat of the liquid."

When all of the substance is melted a further application of heat will again produce rise in temperature and the amount of heat absorbed per pound per degree rise in temperature is known as the "specific heat of the liquid."

When the boiling point of a simple substance is reached,† the further application of heat will not produce any further rise in

* There are some gases which cannot be liquefied by the reduction of temperature alone, high pressure being also a requisite.

† In the case of complex substances, such, for example, as mineral oils, no absolutely fixed boiling points exist. A moderate application of heat to crude oil will drive off one after another of the distillates, such as benzine, gasolene, kerosene, etc., but while the temperature at which each distillate is driven off is different from that of the others, the increase in temperature as distillation progresses is gradual and the distillation temperatures define the product, rather than the product the temperatures.

temperature, but a change of state, vaporization, which will progress at a constant temperature until all is evaporated. The quantity of heat required to evaporate a pound of the liquid is the "latent heat of vaporization" or the "latent heat of the vapor."

When all of the substance is evaporated a further application of heat will again produce a rise in temperature and the quantity of heat absorbed per pound per degree rise in temperature of the vapor is the "specific heat of the vapor." Objection might logically be made to the term "latent heat," since the heat disappearing when solids melt and liquids evaporate really no longer exists as heat, but has been expended in doing work, just as heat units expended in a steam engine are converted into an equivalent amount of work measurable in foot-pounds. A common example of work is the raising of a weight of a certain number of pounds through a certain number of feet of space against the attractive force of gravity. Heat absorbed in the process of melting and evaporation is expended in doing internal work, or in overcoming attractive forces between the molecules of the substance melted or evaporated. The expenditure of heat in separating one molecule from another against the force of molecular attraction, differs in no material sense from that expended in prime movers in separating heavy weights from the earth against the force of gravity.

Heat Units

The standard unit generally adopted in commercial work among English speaking people is the British thermal unit (B.t.u.), which is the equivalent of the amount of heat required to raise the temperature of one pound of water through one degree Fahrenheit at its temperature of maximum density, 39.1° Fahrenheit. This, it may be remembered, Joule found by his historic experiment to be equivalent to 772 foot-pounds of work. More recent experiments, however, have led to the adoption of 778 as a more accurate equivalent. In more purely scientific work the French or metric unit, the calorie, is employed. This unit is the equivalent of the amount of heat required to raise the temperature of one kilogram (2.2+pounds) of water through one degree Centigrade (1.8 degrees Fahrenheit) at its maximum density temperature of 4° Centigrade. By the following experiment, Doctor Joule determined the mechanical equivalent of heat. A paddle fitted to a shaft, which was made to revolve by weights so acting that the exact amount of work done by them could be measured, was placed in a cylindrical vessel containing a definite amount of water at a known temperature. As the paddle revolved the water was agitated and the temperature was found to rise. The energy of the weights had been converted first into motion of the paddle, then into heat in the water. He determined that when mechanical energy is converted into heat the amount of heat produced is proportional to the mechanical energy expended, and specifically, that one calorie represents 424 kilogrammeters of energy. The mechanical equivalent of heat or as it is called Joule's equivalent, is 424 kilogrammeters. In the English system the same equivalent is 772.55 foot-pounds, or one B.t.u., later definitely fixed as 778 foot-pounds.

Since the amount of heat required to raise the temperature of one pound of water one degree Fahrenheit has been made the standard unit by which other quantities of heat are measured, the latent and specific heats above defined are expressed in B.t.u. The specific heat of water is unity.

Since the amount of heat required to raise the temperature of one pound of ice one degree is one-half that of water, the specific heat of water in the solid state is one-half B.t.u., or simply 0.5. Similarly, the latent heat of fusion of water is 144*, the latent heat of vaporization is 966, and the specific heat of steam, according to whether it is taken at constant pressure or constant volume, is 0.480 and 0.346, respectively.

It will be noted that the specific heat at constant pressure is somewhat greater than the specific heat at constant volume. In determining the latter, the gas is not allowed to expand; in other words, it is maintained at a "constant volume." In determining the former, the volume is allowed to increase with the addition of heat just sufficiently to maintain a "constant pressure." The

* At the New York meeting of the American Society of Mechanical Engineers, a committee appointed by that society to suggest a standard tonnage basis for refrigeration proposed as a unit for measuring cooling effect, the equivalent of the heat required to melt one pound of ice, *i.e.*, 144 B.t.u. The unit for a ton of 2000 pounds of ice-melting capacity was then fixed at 288,000 B.t.u., which, since the rating is always expressed in tons per 24 hours, makes a ton duty equivalent to the rate of 12,000 B.t.u. per hour, or 200 B.t.u. per minute.

work done by the gas in expanding is equivalent to the excess of the amount of heat supplied in the case of constant pressure, over that required to produce the same rise in temperature at constant volume.

Heat Absorbing Capacities of Substances

Under a given pressure there are for every substance, definite and fixed temperatures at which that substance will change from one of its three states, the solid, the liquid, or the gaseous, to another; and at each change that substance absorbs or liberates an amount of heat, which, though varying slightly at different temperatures, is more or less a constant and characteristic of that particular change of state of that particular substance. So constant is the temperature at which these changes of state take place that the simple determination of the boiling point under atmospheric pressure, of some of the more common liquids such as alcohol, water, ammonia, etc., is sufficient to establish their identities. At atmospheric pressure alcohol boils at 173° Fahrenheit, water at 212° Fahrenheit and ammonia at −28½° Fahrenheit. Except for the question of the facility of conducting the experiments, the freezing points of the liquids in question might just as well be employed to determine their identity. Absolute alcohol, for example, freezes at −202° Fahrenheit and water at 32° Fahrenheit. In general, the addition of a foreign substance capable of being dissolved in either of these liquids has the effect of lowering the freezing point and raising the boiling point.

Similarly, the amount of heat absorbed or liberated when a fixed quantity of one of the above liquids changes state is so nearly a constant that it could be employed to establish the identity of the liquid were it not for the difficulty of determining the exact amount of heat involved. A pound of water, for example, absorbs 144 B.t.u. in changing from the solid to the liquid state, and 966 B.t.u. in changing from the liquid to the gaseous state. Similarly, in changing from the liquid to the gaseous state a pound of anhydrous ammonia absorbs 555 B.t.u.

Flow of Heat

It has been stated above that there is always a tendency for heat to flow from one body to another wherever there is a difference in temperature. For the present purpose, it may be generally stated that the rate at which the passage of heat takes place is directly proportional to the difference in temperature, and inversely proportional to the amount of resistance offered to its passage by interposed substances. The flow of heat through substances all of which are to a greater or less degree conductors of heat, is analogous to the flow of electricity in an electrical conductor, which, as expressed in Ohm's law, is directly proportional to the voltage or "electromotive force" and inversely proportional to the electrical resistance of the conductor. Expressed as an equation

$$C=\frac{E}{R}, \text{ or (current in amperes)} = \frac{\text{(Electro Motive Force in Volts.)}}{\text{(Resistance in Ohms)}}$$

The flow of heat may also be compared to that of a gas or a liquid in a pipe where the quantity discharged is directly proportional to the difference in pressure and inversely proportional to the friction encountered.

Passage of heat may take place by convection, conduction, or radiation, and in the general case, by all three methods simultaneously, so that the mathematical expression for the total heat transfer between two bodies of different temperatures becomes somewhat complex. Fortunately, in a great number of engineering problems the amount of heat transmitted by conduction is so far in excess of that transmitted by radiation and convection that the last two factors may be ignored entirely or introduced in terms of conduction.

Heat and Cold, Relative Terms

Much of the popular misconception of the art of refrigeration has arisen through our proneness to group temperatures into such technically meaningless classes as "warm," "hot," "cool" and "cold," according to their apparent relation to the widely varying temperatures of our surroundings, which we erroneously come to look upon as a kind of variable zero from which all other temperatures should be measured. This in turn gives rise to the erroneous notion that a substance is capable of heating or refrigerating other substances according to whether it appears hot or cold to the touch.

While this gives a more or less correct idea as to the ability of the substance in question to heat or refrigerate our bodies, it cannot in the general case give any idea of its ability to heat or refrigerate other bodies at widely different temperatures.

If one places his hand in water much higher in temperature than his body it is said to be hot. If he places his other hand in water much lower in temperature than his body, it is said to be cold. If both hands are then placed in water of the same temperature as his body it will feel hot to the hand that was in the cold, and cold to the hand that was in the hot water.

Since heat and cold are relative terms arising from comparison with normal temperatures, a temperature described as being hot in winter might be regarded as being cold in summer. The fact that an object is already cold does not prevent its being made still colder by the further removal of heat. The coldest substances known are still possessed of a large quantity of heat, only part of which can be abstracted by any known method. Could all of the heat be removed from a substance, the resulting temperature would be absolute zero, or about 466° below our present Fahrenheit zero. At this point all chemical action would cease, and neither animal nor vegetable life could exist.

Since refrigeration, which occurs whenever there is a flow of heat from a relatively warmer to a relatively cooler body, may take place at any temperature regardless of whether it is above or below that of our surroundings, the melting of iron in a blast furnace may be said to refrigerate its contents just as truly as the melting of ice does the contents of a refrigerator.

Where the refrigerating effect is due to the direct passage of heat from one substance to another, it is almost axiomatic that the cooling of one can take place only with the equivalent heating of the other. The heating and cooling are the same operation seen from two diametrically opposite viewpoints.

Refrigeration by the Melting of Solids

The most common examples furnished by nature, of processes by which the refrigerating of one body is accomplished by a corresponding heating of another, are the changing of water into ice and its subsequent melting to form water. In the former process, heat passes from the water to the air, the water being refrigerated and the air heated. Conversely in the latter process, heat passes from the air to the ice, the air being refrigerated and the water heated. Both these processes ordinarily take place under atmospheric pressure and at 32° Fahrenheit.

As another example may be cited the congealing and subsequent melting of mercury at −39° Fahrenheit, and that of cast iron at about 2000° Fahrenheit, or in fact that of any fusible substance at its temperature of fusion. In the foregoing examples, absorption of heat, or refrigeration, involves the latent heat of fusion of the substances, water, mercury and iron, in question.

Refrigeration by the Evaporation of Liquids

Another means of bringing about the absorption of heat or refrigeration is by the evaporation of liquids. This involves the latent heat of vaporization, and the fact that the latent heat of vaporization of a substance is greater than the latent heat of fusion is one reason why methods involving the former factor are the more commonly employed in connection with artificial refrigerating systems.

Probably the most common example in nature of refrigeration produced by the evaporation of a liquid, is the cooling effect of summer showers, in which the evaporation of a part of the water precipitated cools the dry, hot air which absorbs it. Another well-known cooling effect is encountered when one sits in a draft after perspiring freely. The effect of the draft is to continually displace the stratum of warm, saturated air lying next the skin by cool, dry air. To evaporate the moisture, heat is abstracted both from the air and from the skin. This continued rapid abstraction of heat from the skin is more rapid than the heat supply from the blood. The temperature of the skin falls so rapidly as to eventually result in congestion and the resultant effect known as a cold. Among less common examples may be cited the method of cooling drinking water sometimes employed on shipboard, *i.e.*, by exposing it to the wind in porous tile vessels, the evaporation of a part of the water through the walls of which refrigerates the portion remaining in the vessel. In India water is actually frozen by the rapid evaporation of part of it exposed in shallow earthen trays to the clear, dry night air.

Refrigerating Temperatures and Pressures

Evaporation can only take place when a fixed quantity of heat is absorbed. Conversely, absorption of heat, or refrigeration, always occurs when a substance is evaporated, whether the evaporation takes place rapidly at the boiling point or slowly at a temperature far below its boiling point. Since no two simple substances boil under the same conditions of temperature, it follows that it may be possible to produce refrigeration either at a given temperature by the evaporation of different substances at different pressures, or within certain limits, to produce refrigeration at quite widely different temperatures by the evaporation of the same substance at different pressures.

In general the most desirable working medium is that substance which has either its latent heat of vaporization or latent heat of fusion available under not too abnormal conditions of temperature and pressure.

While evaporation of water in the tubes of a water-tube boiler refrigerates the furnace gases in almost the same way as the evaporation of other liquids (so-called refrigerating media) does the air surrounding the similar pipe coils in direct-expansion refrigerating systems, the fact that the latent heat of vaporization of steam is not generally available at less than 212° Fahrenheit under atmospheric pressure has up to the present time prevented the extensive use of that medium in connection with what are commonly termed "refrigerating systems."

The availability of the latent heat of vaporization of water at 32° Fahrenheit requires the usually prohibitive vacuum of 29.76 inches of mercury, but the fact that the latent heat of fusion of water is available at 32° Fahrenheit under atmospheric pressure, allows artificial ice to become as important a factor in our general scheme of domestic and commercial economy as natural ice, the freezing and subsequent melting of which in our lakes and rivers protects both animal and vegetable life through the tempering of extreme temperatures, is in Nature's economy.

CHAPTER II

THE DEVELOPMENT OF MECHANICAL REFRIGERATION

SOURCES OF HEAT

IT has already been stated that in the study of mechanical refrigeration the real entlty to which we must direct our attention is heat. The two principal sources of heat are chemical reaction and solar radiation. The most important of all chemical reactions to the engineer, and in fact to the mechanical world, is that of combustion; but even the heat due to combustion has its origin largely in solar radiation.

Primitive man had to depend largely upon solar radiation for his physical comfort. When he enjoyed warm weather, however, his food would not keep, and when the weather was cold enough to keep his food he had to resort to the combustion of the more easily procurable fuels in order to keep warm.

Long before the appearance of man, however, radiant solar energy had been carrying out a process by which carbon dioxide from the air was broken up into free oxygen and fixed carbon in the plant structure of vast vegetable growths, which after becoming buried under the surface of the earth and falling into partial decay, formed our present coal deposits. Radiant solar energy available through chemical reaction, the combustion of coal, now supplies the greater part of the power required to move our commercial world.

While man's industry in digging in the ground for stored heat enabled him to solve the problem of keeping warm, it remained for his ingenuity to devise methods for preserving his food—first by the direct use of the natural cooling medium, ice, and later, as his creative ability increased by the indirect use of artificial cold produced by the expenditure of the natural heating material, coal.

Man's first lesson in refrigeration taught him that water solidified by some mysterious property possessed by the cold north wind could be preserved to serve him during the summer, if carefully stored away in caves in the earth. With certain minor refinements in the operations, this same crude method employed by our primeval ancestors is still being used to no inconsiderable extent to-day. The fact, however, that summer requirements must be anticipated at least one season by both Nature and man, has combined with a score of other elements to force the introduction of more scientific methods.

Frigorific Mixtures

Laboratory methods of producing low temperature by means of so-called "frigorific mixtures," by which a perceptible drop in temperature is produced by certain endothermic chemical reactions and solutions have been known for at least three centuries.

Probably the most common example of a frigorific mixture is that of ice or snow and salt. The addition of a foreign substance to a liquid lowers its freezing point. The effect of the addition of different amounts of common salt (NaCl) and that of calcium chloride (CaCl) is clearly set forth in the table on page 145. Since the addition of 10 per cent. of salt, by weight, to water lowers its freezing point to 18.7° Fahrenheit and prevents its changing to the solid state till that temperature is reached, it would follow that the addition of the same percentage of salt to snow or finely divided ice would cause it to return to the liquid state at all temperatures above 18.7° Fahrenheit. The result is that the ice at once begins to melt, but to do so it must absorb 144 B.t.u. of heat per pound, and in the event that this heat is not forthcoming, the temperature of the mixture will continue to fall until the freezing point corresponding to the per cent. solution is reached. At this point it will continue to exist in the solid state, and aside from the lesser intimacy of the mixture of the two constituents, ice and salt, will be the same substance as frozen brine of the same per cent. composition, and will exist under the same conditions of temperature.

Heat Absorbing Capacities of Substances

The heat absorbing capacity of a substance when only its temperature is raised, and its state remains the same, is comparatively small. That of water is 1 B.t.u. per degree rise in temperature, in the liquid state, and less in both the solid and gaseous states. The specific heat of other substances in all three states is in the general case less than unity or that of water in the liquid state.

The heat absorbing capacity of a substance available in its change of state, involving its *latent* heat of fusion and vaporization, is comparatively large. That of water is 144 B.t.u. for fusion, and about 966 B.t.u. for evaporation.

In general, the greatest heat absorbing capacity of any substance is in its latent heat of vaporization available at its boiling point.

Since increasing the pressure has the effect of raising the boiling point, or the temperature at which the liquid vaporizes, and decreasing the pressure has the effect of lowering it, it is only natural that reduction in pressure below that of the atmosphere (or vacuum) was first employed in attempts to cause some of the better known liquids to boil at sufficiently low temperatures to produce artificial "cold."

Early Experimenters

While there is evidence that some little experimenting was done with liquids under vacuum as early as 1755 there is no authentic information that anything of real importance was accomplished until the early part of the last century, when several independent inventors built experimental machines, none of which, however, seem to have produced any very encouraging results until Jacob Perkins, an Englishman, developed an ether-compression machine which he patented in 1834.

Perkins operated his ether-compression machine under vacuum, much as our present-day ammonia-compression machines are operated when low temperatures are required. His patent of August, 1834, describes his machine as being composed of the four principal parts which constitute our present machines, viz., a containing chamber or evaporator in which the refrigerating medium evaporates, and through the walls of which it absorbs heat from the substance it is desired to refrigerate; a pump or compressor for drawing the evaporated refrigerant from its containing chamber and exerting upon it sufficient pressure to cause it to liquefy when cooled; the cooler or condenser consisting of a coil of pipes submerged in water, and in which the refrigerant is cooled and liquefied after compression, and a regulating or expansion valve for controlling the flow of the refrigerant liquefied in the condenser as it passes under the higher condensing pressure to the evaporator maintained under a lower pressure due to the action of the compressor.

The evaporator shown in the Perkins patent consists of a circular tank made of two dished metallic disks, which receptacle was submerged in the fluid to be refrigerated. The compressor was designed to be operated by hand, but otherwise the general arrangement of its working parts, except for the fact that the cylinder was inverted, was very much the same as that of our present small single-acting ammonia compressors with both suction and discharge valves in the head. A simple submerged condenser, consisting of a single zigzag coil, and a hand-operated expansion valve were employed.

Both the compression and the absorption machines find their origin in the demonstrated possibility of liquefying so-called gases. In 1823 Faraday announced to the world that he had succeeded in liquefying chlorine, ammonia and carbon-dioxide, as well as several other gases of less importance to the refrigerating industry.

By the same method by which Faraday first accidentally liquefied chlorine, carbon-dioxide, ammonia, sulphur-dioxide, methyl-ether, Pictet fluid, sulphuric ether, ethyl chloride, water, and other substances may be liquefied experimentally under the proper conditions of temperature and pressure.

The first ammonia-absorption machine recognized as such was invented by Carré about the year 1855, in which same year Harrison, an Australian, and Professor Twining, an American, are said to have independently perfected the Perkins ether machine, the latter inventor having performed the then marvelous feat of artificially freezing blocks of ice with fish inside.

Contributory Factors

Not the least among the influences that have combined to forward the development of mechanical refrigeration has been the pollution by sewage of the water from which the natural ice crops are harvested. This has been especially true since it has been demonstrated that negative bacteriological tests, no matter how carefully conducted, are not insurance against typhoid, and there is accordingly well-founded and rapidly growing prejudice against all ice known to be cut from sewage-bearing streams and lakes.

The breweries were among the first to adopt improved methods of cooling, largely because of contamination of products through the unsanitary conditions inevitably resulting from the use of ice.

This, together with the heavy labor charge which it entails, forced the development of a system of mechanical refrigeration. While the item of contamination was somewhat less important in the abattoirs than in the breweries, the enormous amounts of ice consumed and the uncertainty of the natural crop, made the adoption of mechanical refrigeration in this industry imperative from an economic standpoint.

In cold-storage work the inability of ice, without the addition of salt, to produce sufficiently low temperatures to satisfactorily preserve many perishable products, has of late years, at least, probably been the most potent factor in the combination to force the substitution of mechanical systems for ice in the cold-storage industry.

The demand for artificial means for producing refrigeration having first appeared among the larger industries above cited, the builders of refrigerating machines first set about to supply systems best adapted to their peculiar requirements. Later as the requirements of the larger consumers of cold began to show signs of having limitations, the builders began to look for other fields and quite naturally began to develop systems better adapted to the requirements of smaller consumers of ice. While the idea is still current with a certain class of consumers that food products kept in mechanically cooled compartments suffer thereby, the intelligent merchant is willing to acknowledge the superiority of mechanical cooling means in almost every case.

Types of Small Machines

The chief factor which fixes the capacity limit under which it becomes commercially impracticable to install small mechanical refrigerating plants has been the cost of attendance which is practically as great for all smaller sizes as for those of from five to ten tons capacity, and this, in the general case in which steam power is employed 24 hours per day, solely for the operation of the refrigeration plant, becomes practically prohibitive. The first step toward surmounting this obstacle was to operate the plant in the daytime only in order to eliminate the expense of night attendance. In order to do this the brine-circulating system commonly called the "brine system," which will be described in detail later, was introduced. The brine system, however, usually consumes power 24 hours per day, and in order to reduce this expense the "congealing-tank" system, which will also be described in detail later, is employed to some considerable extent.

Method of Operation

Where steam is not required for other purposes and the amount of power necessary for the refrigerating plant is small, the substitution of combustion-engine power is often capable of reducing the cost of fuel as well as that of attendance; the former where cheap kerosene, crude or fuel oils of light density can be procured, and the latter where licensed attendants, insurance, and other factors work disadvantageously for the small steam plant.

The comparatively low cost of power produced by combustion engines, coupled with the slight expense for attendance now made possible by the application of certain automatic regulating and safety devices, makes this type of plant, when properly installed either with or without "congealing tanks" as the case may require, a most practical and satisfactory plant for the small consumers of ice. To the end of producing a small mechanical refrigerating plant capable of operating with still less attendance, so-called completely automatic systems which will also be described in detail, have been developed. While it is a fallacy to suppose that these machines will operate without some attendance, they are often capable of operation with far less attendance than any other type, and were it not for the usually comparatively high cost of electric power necessary for the operation of a completely automatic plant, the advantages gained would undoubtedly more than compensate for the high first cost of such plants.

No one type of plant can be expected to be the most advantageous under all requirements, and the relative advantages and disadvantages of one over the other in first cost, operating cost, superiority of design, and safety should be carefully considered and balanced up before a selection is made.

CHAPTER III

COMMERCIAL SYSTEMS OF REFRIGERATION

Absorption and Compression System

Practical mechanical refrigeration may be said to date back to 1855, in which year the ammonia-absorption machine was invented by Carré, and the Perkins ether-compression machine, patented in 1834, was simultaneously commercialized by two different inventors in two different countries. It is a rather strange coincidence that the development of the two systems so commonly employed to-day, *i. e.*, the absorption and the compression systems, should have been commenced the same year.

Generally speaking, while both the absorption and the compression systems have been employed in connection with both the direct expansion and the brine systems for the production of both medium and extremely low temperatures, brine has been used more commonly in connection with the absorption machine and direct expansion in connection with the compression machine. The former has been employed more frequently for extremely low and the latter for comparatively low temperatures. The absorption machine is generally conceded to be more economical than the compression machine when producing extremely low temperatures, under which condition, because of the lightness of the gas under the correspondingly low pressures, the compression machine loses greatly in both efficiency and capacity. At the present time, however, it is problematical whether the substitution of the higher-efficiency gas engine as a prime mover for the inherently low-efficiency steam engine will not put the compression system in position to produce as economical results as the absorption system, even under the disadvantage of low temperatures.

Since the compression system is in far more common use today, and since a large number of the parts of the two systems are common (see Fig. 22), the details of construction of the compression machine will first be considered.

The function of a refrigerating means, whether it be an absorption or a compression machine, or simply a bunker full of ice, is to provide a heat-absorbing medium which, after it has absorbed its fill of heat from the products to be cooled in the cold-storage rooms, may be removed from the coolers so that the heat absorbed may also be removed. After its removal from the coolers, this medium may be divested of its heat, after which it may be allowed to return to the coolers to absorb more heat, as in the case of ammonia or brine circulated through coolers; or it may be thrown away and a new supply introduced, as in the case of cooling by ice.

Ice Bunker System

Bunkers for ice, or other cooling means, are usually constructed in the form of a tank with one side removed, as shown in Fig. 1, amples paces for ducts being allowed between the sides of the tank and the sides of the room to permit the air to circulate freely. The ice in this case is stored on a water-tight floor over the compartment to be cooled, the cooling being effected by a natural circulation of air up over the ice and down through the cold-storage compartment.

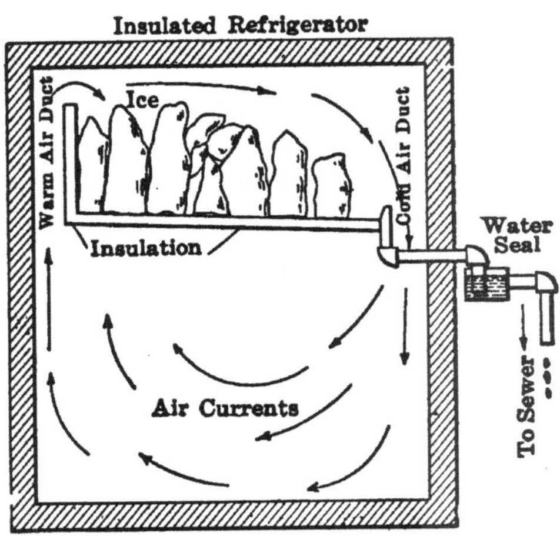

Fig. 1.—Refrigerator Cooled by Ice Bunker

Were the ice stored in a bunker of the form of a tank without one side removed it is obvious that the heavy cold air, like water, would sink to the bottom of the tank and there would be little or no circulation. If, on the other hand, both sides of the tank are removed it is equally obvious that the tendency would be for the cold air to flow off the ice in both directions. This would give rise to conflicting currents of air which would check the circulation, and for the same rate of air circulation would require a greater difference in temperature between the air in the bunker and that in the cooler below.

Natural circulation of air in a cold-storage compartment may be compared to natural draft in a chimney except that the purpose and the causes are reversed. In the case of natural draft, air and gases are artificially heated in a furnace. Since the heat causes a given volume to expand to a greater volume, its weight per unit of volume is decreased, causing it to rise in the cooler, heavier air above, just as a piece of cork or any comparatively light substance rises in water.

In the case of natural circulation of air in a cold-storage compartment the air is artificially cooled in a bunker containing ice, or some other cooling means. Since the extraction of heat, or cooling, causes a given volume to contract to a lesser volume, its weight per unit of volume is increased, causing it to sink in the lighter air of the cooler below, just as a stone or any heavy substance sinks in water.

Since both the bunker and the room below are completely filled with air the sinking of the cold air on one side can take place only by the displacement of an equal volume of warm air, which rises and passes into the bunker on the other side. The flow of air known as "draft" in a chimney, "air circulation" in a cold storage compartment, or "wind" on the surface of the earth, takes place only because of a difference in pressures due to a difference in temperature.* To ensure this maximum "draft" or "air circulation" it is necessary to maintain a liberal difference in temperature and a reasonable "head." In the case of chimneys this is accomplished by so constructing the walls as to prevent undue radiation of heat and of such a height as to give a difference in weight, equal to the required draft, between the hot air within and the cold air without the chimney. Similarly, in the case of the

* Since the difference in temperature is greater, production of draft in a chimney furnishes the better example.

The difference in weight per cubic foot between air at 62° Fahrenheit and 500° Fahrenheit is $0.0761-0.0413=0.0348$ pound. A column of 62° air 100 feet high would accordingly weigh 7.61 pounds, and an equal column of 500° air, 4.13 pounds, making a difference in weight of $7.61-4.13=3.48$ pounds. If the column of 500° air be in a chimney 100 feet high, and the 62° air outside, the actual difference in pressure tending to produce upward flow of air in the chimney will be 3.48 pounds per square foot, or $16\times 3.48\div 144=0.3866$ ounces per square inch. Chimney draft is usually expressed in inches of water. Water at 62° Fahrenheit weighs 62.32 pounds per cubic foot. The pressure due to a column of water 1 foot high is accordingly 62.32 pounds per square foot. The pressure in ounces per square inch due to a column of water one inch high is $62.32\times 16\div(144\times 12)=0.577$. The difference in pressure of 0.3866 ounces per square inch, due to the 100-foot stack and a difference in temperature of 500–62° Fahrenheit would accordingly be equivalent to $0.3866\div 0.577=0.67$ inches of water; or since the specific gravity of mercury is 13.58 (13.58 times the weight of water) the draft as expressed in inches of mercury would be $0.67\div 13.58=0.049$.

cold-storage bunker the uptakes should be insulated to prevent the cooling of the rising column of warm air and the height of the cooler compared to the width between hot and cold air ducts should be kept as great as possible.

Bunker Insulation

To prevent the cooling of the air lying next to the floor in the bunker, which not only retards circulation by reducing the difference in temperature, but may also precipitate moisture when the warm air approaches saturation, the bunker floors should also be insulated. When operating with properly insulated bunkers and liberal ducts, the circulation should be sufficiently rapid to carry away any moisture-saturated air that may enter the cooler from the outside or that may have become saturated through contact with stored products, to the ice chamber before its moisture can be precipitated by contact with the cold surfaces of the cooler. When cooled in this way the excess moisture is precipitated in the ice bunker and flows away with the melted ice.

Where the warm air ducts or the bunker floors are not insulated, excess dampness can be prevented only by the use of some such deliquescent salt as calcium chloride. This is the case when the air of the cooler is not allowed to come in contact with the melting ice but is chilled by contact with the metallic floors of overhead bunkers or the walls of tank bunkers.

Where salt is used so that the resulting temperatures are below freezing, the precipitated moisture is frozen on the heat-absorbing surfaces and the difficulties arising from condensation are avoided, but in time the accumulated ice has to be removed, giving rise to other difficulties.

Exceptionally well insulated coolers have been constructed in which it is possible to maintain temperatures as low as from 38° to 40° Fahrenheit in compartments cooled by natural circulation and ice without the use of salt. The average cooler, however, is not capable of producing such favorable temperatures, and the present-day demand for lower temperatures makes it necessary to resort to other means. The addition of salt allows the natural circulation and ice system to satisfy a few such requirements, and the further addition of a fan to force the circulation may allow it to include a few more.

Gravity Brine System

For still further extending the application of ice cooling, the system illustrated in Fig. 2 has been devised. This system consists essentially of a tank for holding the ice and salt in the proportion required to produce the desired temperatures, and a continuous pipe circuit, a part of which is located in the cold-storage compartment, where it absorbs heat, and the other part in the ice tank, where it gives up heat to the ice or freezing mixture of ice and salt, as the case may be. The pipe is filled with brine of just

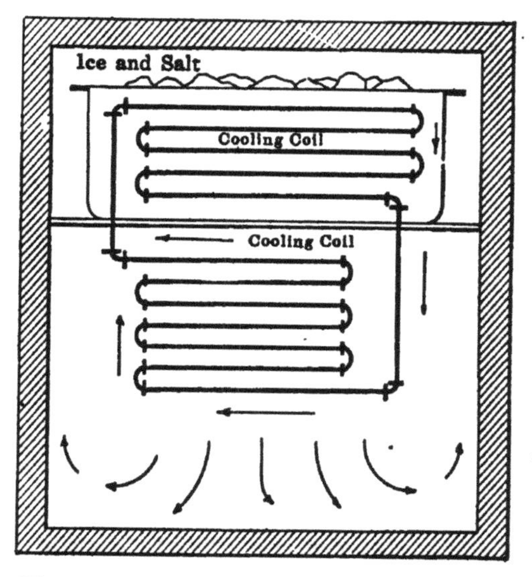

Fig. 2.—Refrigerator Cooled by Gravity Brine System

sufficient density to insure against freezing. The brine in the coil of pipe located in the tank, becoming heavier as it is cooled by the surrounding ice and salt, flows down into the coil located in the cold-storage compartment, causing the warmer, lighter brine to pass upward and take its place in the ice tank. Heat is conveyed from the cold-storage compartment to the ice chamber by the natural circulation of the conveying medium as in the preceding case, but in this case the medium is a liquid, brine, while in the preceding one it was a gas, air. An abnormal rise in temperature of the cooler increases the velocity of the brine circulation and consequently increases the refrigerating capacity of the system. By the use of this system it is possible to prevent the air of the cold-storage compartment from becoming contaminated by contact with melting ice and unsanitary ice bunkers.

ELEMENTARY MECHANICAL SYSTEMS—DIRECT EXPANSION

When the refrigerating fluid is a condensable gas, or, more accurately speaking, a liquid having a sufficiently low boiling point to evaporate and absorb heat under conveniently low pressures to produce the desired temperatures the process of mechanical refrigeration may be explained as follows:

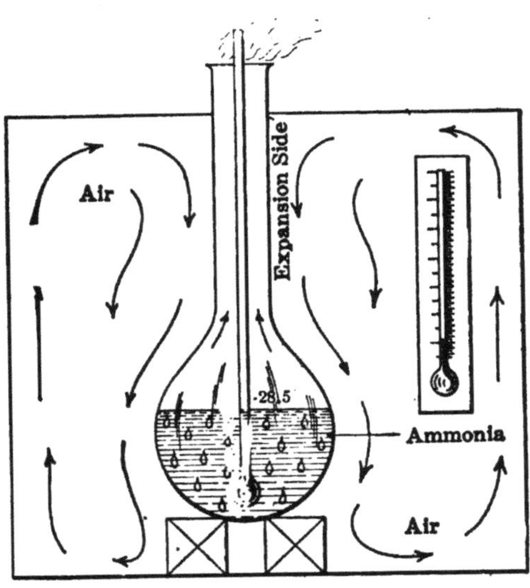

Fig. 3.—Elementary Direct-Expansion System

Fig. 3 represents a flask partly filled with a refrigerant such as anhydrous ammonia. Since the pressure on the refrigerant in the open flask is only that of the atmosphere, it will boil at—28.5° Fahrenheit. As this temperature is far below that of the surrounding air under usual conditions, heat will pass from the air into the refrigerant; the former will be refrigerated and the latter heated up to the boiling point, if it is not already boiling. At the boiling point the absorption of a definite amount of heat from the air effects the evaporation of a definite amount of the liquid; the anhydrous ammonia, absorbing heat directly through the walls of the containing vessel from the surrounding air, operates just as it and similar refrigerating media do in the direct-expansion refrigerating system. This cooling effect is hastened by a marked circulation of air around the flask, occasioned by the greater specific gravity or weight of the cooled film of air lying next to the flask, which flows down and away at the bottom, causing the warmer air to rise and take its place at the top.

Brine Circulation System

The brine system has its elementary counterpart in such an arrangement as is illustrated in Fig. 4, in which the flask containing the boiling anhydrous ammonia is immersed in a second vessel, such as a large beaker, filled with a brine solution of sodium

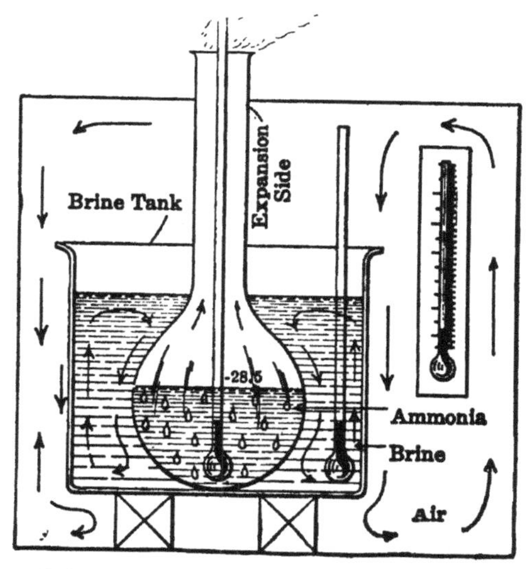

Fig. 4.—Elementary Brine System

chloride (NaCl), calcium chloride (CaCl), or other salt in water. The low first cost, together with certain physical and chemical characteristics, has practically limited the commercial brine system to the use of either sodium or calcium chloride solutions. As in the preceding case, the boiling refrigerating medium absorbs heat from the surrounding medium, in this case brine, which we will assume is of such a nature as not to freeze at— 28.5° Fahrenheit. Since this solution does not freeze at the temperature of the evaporating refrigerant, and the latent heat of fusion of the liquid is not extracted, an evaporation of only a comparatively small amount of the refrigerant suffices to cool the surrounding solution almost down to −28.5° Fahrenheit. This cooling effect is hastened by a marked circulation within the solution itself, set up by the difference in specific gravity of the colder part of the liquid lying next to the flask, which sinks to the bottom of the beaker, causing the warmer part to flow in and take its place at the top.

The brine in the beaker having become colder than the surrounding air, heat flows from the air to the brine, just as it flowed from the air to the ammonia in the preceding case. It will be noted, however, that difference in temperature, and accordingly the flow of heat, between the air and the brine can never be so rapid as that between the air and the ammonia, except in the limiting case in which the brine and the ammonia are of the same temperature, when, unfortunately, the ammonia would have no cooling effect on the brine. If the same amount of the same refrigerant be placed in flasks of the same shape and size in both the foregoing experiments, it will be found that the brine employed in the second case will assume a temperature intermediate between that of the boiling ammonia and that of the surrounding air, and were there any way to measure the cooling effect produced on the air, it would probably be found that less heat would be absorbed in the latter than in the former case, notwithstanding the fact that the heat-absorbing surface of the beaker is greater than that of the flask. A very crude idea of this difference might be gained by surrounding the vessels used in each experiment by boxes made of the same size and material and noting the temperature of the air within.

The flasks containing the ammonia employed in the above examples correspond to the expansion pipes immersed in and employed for the purpose of cooling the brine in the brine system described below, and the beaker containing the brine, to the brine pipes immersed in and employed for cooling the air in the cold-storage compartments.

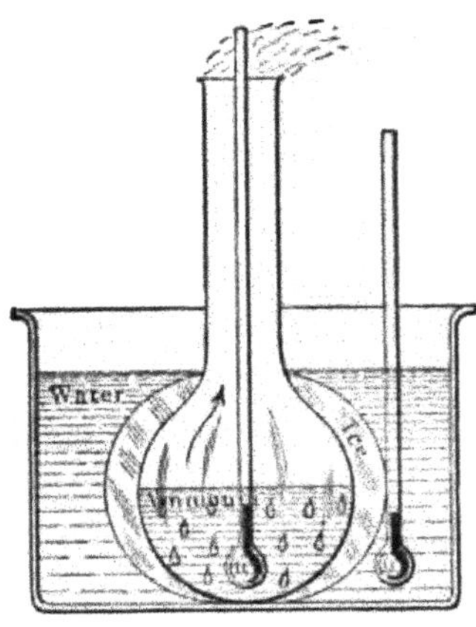

Fig. 5.—Elementary Ice Freezing System

Ice Freezing System

If the flask of ammonia employed in the second experiment had been immersed in water, as shown in Fig. 5, instead of brine, as shown in Fig. 4, the water which freezes under atmospheric pressure at 32° Fahrenheit, will give up first its specific heat in cooling, then its latent heat in freezing, to the anhydrous ammonia boiling under atmospheric pressure at −28.5° Fahrenheit, with the result that the water will first be cooled from its initial temperature to 32° Fahrenheit, then be frozen at 32°. If sufficient ammonia still remains and heat is not absorbed too rapidly from the surrounding air, the ice will be cooled to several degrees below 32°; in fact, that part lying nearest the flask may be chilled to the temperature of the evaporating liquid, −28.5° Fahrenheit.*

Commercially Practical System

Were it not for the initial cost of the refrigerating medium, this elementary refrigerating system in which the ammonia is allowed to escape to the atmosphere after evaporation might find commercial application, but since anhydrous ammonia commands a practically fixed market price of $0.25 and upward per pound, according to the distance from point of production, such a system would be eminently impracticable.

In order, therefore, to make the systems illustrated in Figs. 3, 4, 5, and 6 commercially practicable, means must be provided for converting the gasified anhydrous ammonia back into the liquid state. A commercial refrigerating system is simply a convenient mechanical means for circulating a heat-absorbing medium through the cooler, and of removing from this medium the heat absorbed *en route*. Heat is absorbed in the cooler from the atmosphere, products, etc., by virtue of the fact that the refrigerating medium is lower in temperature. This absorption of heat brings about evaporation and before the refrigerant can be made to do

*If a pound of anhydrous ammonia be evaporated from and at −28.5° Fahrenheit in the flask, the amount of heat which it will absorb will be 573 B.t.u., which is its latent heat of vaporization. If now it be assumed that this heat is all drawn from three pounds of water at an initial temperature of 79° Fahrenheit, it will just suffice, assuming no losses, to cool it down to the freezing point and then to freeze it at 32° Fahrenheit, 47 B.t.u. for each of the three pounds being required to overcome the specific heat of the water in cooling it through 47 degrees, and 144 B.t.u. for each pound of water at 32° Fahrenheit being required to overcome the latent heat of fusion.

Three pounds water × drop in temperature from 79 to 32 degrees × specific heat of water (unity) + three pounds water × latent heat of fusion of water (144 B.t.u.) = latent heat of vaporization of the ammonia (573 B.t.u.) required to cool and freeze the three pounds of water.

$$3 \times 1(79-32) + 3 \times 144 = 573.$$

more cooling work it must be again liquefied. In other words, after the ammonia has evaporated and has absorbed its full complement of heat, much as the sponge sucks up its fill of water, the heat and water with which the ammonia and a sponge are respectively charged must be squeezed out before they can again perform the function of absorption. In the case of the sponge, this is accomplished by the simple application of pressure. In the case of

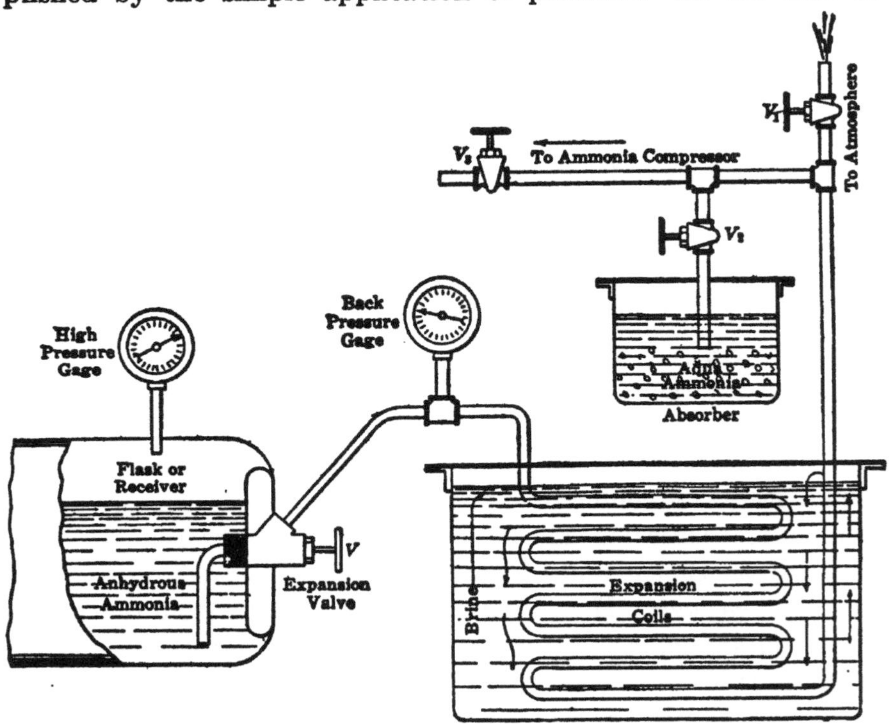

Fig. 6.—Expansion Side of Elementary Brine System

ammonia, pressure must not only be applied, but, since heat can be made to flow only from a relatively warmer to a relatively cooler substance, some cooling medium must be provided as well. The two cheapest and otherwise most convenient natural cooling media we have at our command are air and water. On account of the low specific heat of the former and the fact that it is usually at a higher temperature than water found in the same locality, we are practically limited to the use of water.

If carried out in an ice plant instead of a laboratory, the apparatus requisite for the above experiments would probably appear more nearly as in Fig. 6. In this illustration the glass flask is replaced by an iron flask, or drum, such as is commonly used for shipping ammonia, carbon-dioxide, and other liquefied gases. The liquid from this flask is allowed to "expand" or escape in a small stream through a valve *V* and a pipe leading to an expansion coil. After traversing this coil, the vaporized ammonia escapes to the atmosphere. This outlet, however, is provided with a valve *V*, which, when closed, diverts the ammonia through a second pipe with two outlets, each provided with a valve, the one leading to a tank of water and the other to an ammonia compressor. If diverted into the water, the ammonia vapor will be absorbed and be recoverable in the form of aqua ammonia, a principle which, as will be shown later, is employed in the absorption type of refrigerating machine. If conducted to an ammonia compressor and suitable condenser, it may be condensed and recovered in the form of anhydrous ammonia, a principle which is employed in the compression type of refrigerating system. The several parts used in commercial refrigerating systems to take the place of those illustrated in Figs. 3, 4 and 5 are represented diagrammatically in the figures that are to follow. Fig. 6*b* is such a conventional representation of an ammonia flask or receiver, a controlling or expansion valve *V* and a container of the boiling liquid ammonia, or expansion coil, which may be connected to any type of device for converting the vaporized refrigerant back into the liquid state. Before this can be accomplished, however, the heat absorbed in the cold-storage rooms must be squeezed out, or rather be induced to flow out, of the refrigerating medium by introducing a secondary medium materially lower in temperature. Since there is no secondary cooling medium available of a lower temperature than the refrigerating medium, even at the temperature at which it returns from the cooler after having absorbed large quantities of heat, it becomes necessary to raise the temperature, or thermal level, of the heat in the refrigerating medium above that of the cooling medium sufficiently to allow it to gravitate into the cooling water. Were the water sufficiently cold, *i.e.*, colder than the gaseous

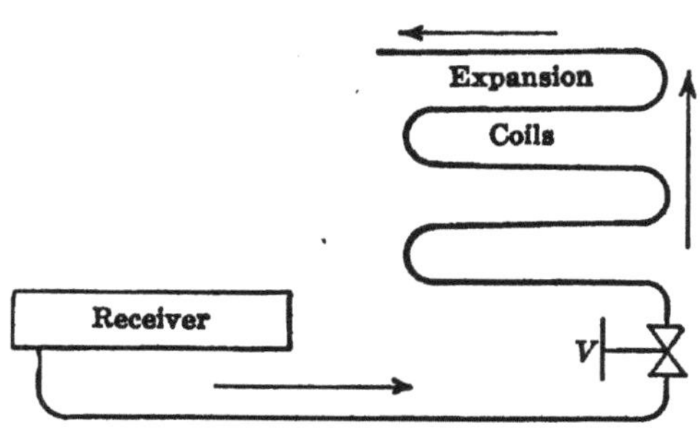

Fig. 6*b*.—Conventionalized Diagram of Expansion Side of Elementary Brine System

ammonia returning from the cooler, the flow would take place without the increase in pressure and temperature, and there would be no need for a compressor. On the other hand, if so cold a medium were available, it would be used directly for absorbing heat in the place of the ammonia or the ammonia and brine, as it would then be sufficiently cold to induce a flow of heat direct from the products to be refrigerated.

ESSENTIAL MEMBERS OF COMMERCIAL SYSTEMS

A commercial refrigerating system consists of a set of pipes, or other containing vessels, in the cooler, in which the refrigerating medium absorbs heat at a low temperature from the products to be refrigerated; a second set of pipes, or other containing vessel, outside of the cooler in which the refrigerating medium gives up its heat to a secondary cooling medium, such as water or air, at a comparatively high temperature; and a compressor or generator.

The pipes located in the cooler, through the walls of which the ammonia absorbs heat from the objects to be refrigerated, are erroneously called expansion coils, and those through the walls of which the ammonia gives up its heat to the natural cooling medium, water, condenser coils.

Since the pressure and consequent temperature of the cold ammonia gas returning from the expansion pipes must be raised before a heat transfer can be made to take place between it and cooling water at ordinary temperatures, the use of a compressor or suitable gas pump, in the case of the compression system, and some other means of heating and increasing the pressure of the refrigerating medium, in case of the absorption system, must be employed. In either system there must also be provided a suitable cooling chamber or condenser in which the cooling water can be brought in sufficiently close proximity to the refrigerating medium to allow the necessary heat flow from the hot ammonia to the cold water to take place.

In practice, this operation is effected as follows: Ammonia, for example, boils, evaporates, and liquefies under a pressure of 16 pounds at a temperature of 0°. In changing from the liquid to the gaseous state it absorbs, and from the gaseous to the liquid state, it gives up an amount of heat equivalent to its latent heat of vaporization, for it must be understood that liquefaction of a vapor at the boiling point of the substance may be effected just as readily by cooling as vaporization can by heating, the only difference in conditions being that it is necessary to provide a colder medium to absorb heat from the gas in the latter case, whereas a warmer one was employed to supply heat to the liquid in the former case. This reversible operation of vaporization and liquefaction may in the general case take place at any temperature and its corresponding pressure.

As there is no natural cooling medium cold enough to liquefy a refrigerating medium such as ammonia gas, for example, at the temperature at which it comes from the cooler, its temperature must be increased. As water is a natural cooling medium, the temperature of the refrigerant is usually raised to a few degrees above the temperature of that medium. In other words, to effect the extracting of heat from the refrigerant by air or water, it is necessary to shift the boiling point of the medium from that of the cooler to some temperature a few degrees higher than that of the condenser. This is the purpose of the generator of the absorption, and the compressor of the compression system. By the direct application of heat in the former and power in the latter, the pressure and temperature of the refrigerant vapor is increased just as the pressure of steam is increased by the application of heat under a boiler and by compression in a steam cylinder at the end of the exhaust stroke.

If the temperature of the available cooling water be 80° it will be necessary to raise the temperature of the vapor to say 100° in order to provide a sufficient difference in temperature to carry out the cooling effect.

In the case of anhydrous ammonia above cited, the original and final temperatures and corresponding pressures are as follows:

Cooler Temp., 0°;	Cooler or "Back" Pres.,	16 lbs.
Condenser Temp., 100°;	Condenser or "Head" Pres.,	200 lbs.

Working Mediums

In the operation of a compression system, almost any gaseous working medium might be employed. In practice, however, the list is limited to only such gases as are capable of being liquefied under ordinary natural temperatures and not to high mechanically produced pressures. Judging from the relative number of commercial installations employing the different media, one may assume that under the average conditions anhydrous ammonia comes nearer to fulfilling all the requirements of a practical working medium than any other.

The system described and illustrated in Fig. 3 consists essentially of a single member, *i.e.*, a containing vessel for the working medium. Here there is no outer containing vessel and no second liquid. In this case the heat passes from the air surrounding the flask, directly to the ammonia, just as the air of a cold-storage compartment is cooled by "direct expansion," a system which is differentiated from the brine system by the

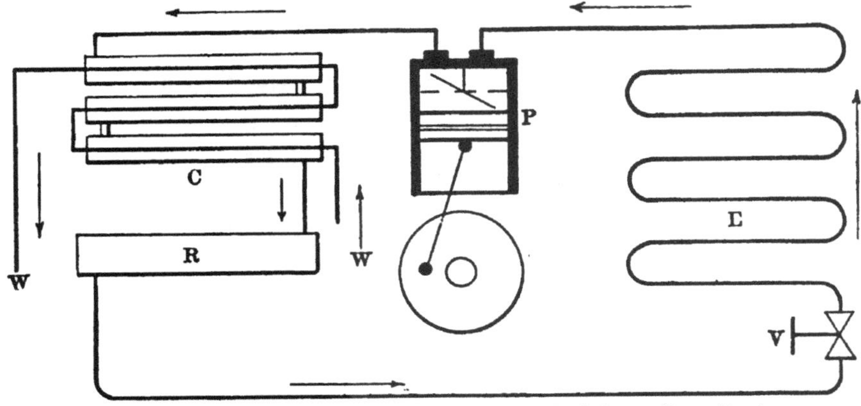

Fig. 7.—Direct-Expansion—Compression Refrigerating System

location of the "expansion coils" containing the boiling or expanding ammonia, in direct contact with the atmosphere of the cold-storage rooms.

The commercial system corresponding to the one illustrated in Fig. 4, in which the evaporation of the ammonia cools the surrounding water, which in turn cools the surrounding air, is the "brine system," which takes its name from the fact that the cooling effect of the refrigerating medium is expended on a more or less uncongealable solution, such as sodium chloride (NaCl) or calcium chloride (CaCl) brine, in which the cooling solution circulated through a secondary system of cools in the cold-storage compartments is made the vehicle for conveying the refrigeration to, or, more properly speaking, the heat away from, the atmosphere in the cold-storage rooms.

The Direct-expansion Compression System

Fig. 7 is a diagrammatic representation of the essential members of a complete compression refrigerating system. *E* represents the direct-expansion coil in which the working medium is evaporated as in the flask, *P* the compressor or pump for increasing the pressure of the gasified ammonia, *C* the condenser for cooling and liquefying the gasified ammonia, and *V* a throttling valve by which the flow of liquefied ammonia under condenser pressure is controlled as it flows from the receiver *R* to the expansion coil *E*, in which a materially lower pressure is maintained by the pump in order that the refrigerating medium may boil at a sufficiently low temperature to absorb heat from and consequently refrigerate the surrounding air, which is already cold. In practice, the systems are somewhat more elaborate, Fig. 15 being a more accurate representation of a commercial system as applied to the direct cooling of cold-storage compartments.

Brine Circulating System

In order that some work of refrigeration may be carried on while the plant proper is shut down either because of accident or

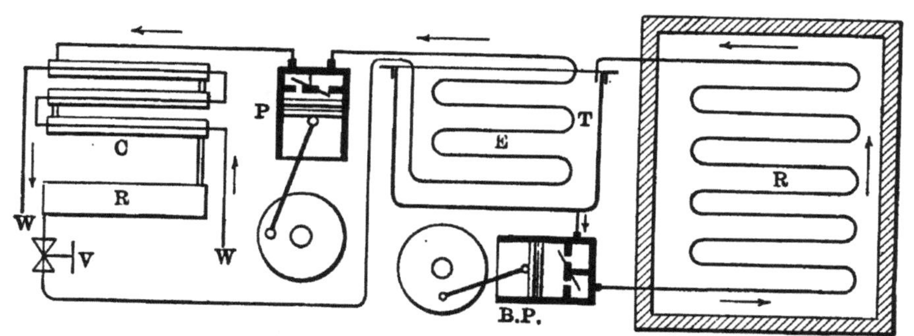

Fig. 8.—Brine-Circulation—Compression Refrigerating System

in order to avoid the expense of skilled attendance during a part of the twenty-four hours, the brine-circulating system is employed. Such a system is represented in the conventionalized diagram, Fig. 8. In addition to the usual members of the direct-expansion refrigerating system represented in Fig. 1, the brine system employs a brine tank *T*, a series of air-cooling or brine coils and a brine pump *B. P.* In this case the initial cooling effect of the evaporating ammonia is expended in the brine which is circulated by the brine pump through the air-cooling coils installed in the cold-storage room *R*.

Salt, either sodium chloride (NaCl) or calcium chloride (CaCl) is required for making the brine, and suitable insulation must be provided for the brine tank, the brine cylinders of the pump and all brine piping outside of the cold-storage compartments in order to reduce to a minimum the losses due to the radiation of cold, or more accurately speaking, the absorption of heat. In addition to the first cost of the above-mentioned items, the operating expense or power required to operate the brine pump must be considered; first, for the circulation of brine from the brine tank through the cooling coils back to the tank; second, to overcome the friction of the brine in traversing the above cycle; third, to produce sufficient additional refrigeration to make up for the heating effect produced mechanically by the circulation of brine and the heat actually absorbed through the brine tank and brine-piping insulation; and fourth, to make up for the greatly reduced efficiency occasioned by the necessity of operating the refrigerating plant proper at a materially lower back pressure in order to produce a sufficiently lower ammonia temperature to make up for the second heat transfer encountered between the atmosphere of the cold-storage compartments and the ammonia gas. The increased operating expense directly resulting from the inherent low efficiency of the brine system is sometimes as high as 25 per cent., and isolated instances have been noted in small plants where the cost of circulating the brine alone was almost as great as that of cooling it.

Congealing Tank System

Instead of the brine system some builders employ what is known as the "congealing-tank" system, represented diagrammatically in Fig. 9. The first cost of the brine pump and the power required to operate it, and the cost of brine pipe and tank insulation, as well as the losses through the same, are avoided in this system by virtually splitting up the main brine tank and installing the pieces commonly called "congealing tank" in the several cold-storage compartments. In this case the heat-absorbing surface of the several smaller tanks entirely replaces that of the brine coils. Here, however, the desirability from a mechanical standpoint of making the tanks as small as possible debars the carrying of sufficiently large volumes of brine to provide for refrigerating the rooms for any great length of time by the rise in temperature only of the brine. The amount of refrigeration performed measured in B.t.u. would be the product of the pounds of brine, the number of degrees rise in temperature and the specific heat of the brine:

$$B.t.u. = lb.\ brine \times specific\ heat\ of\ brine \times (t - t_1).$$

In order to store more refrigeration in the comparatively small volume of the congealing tanks a weak solution of brine is employed, which, in freezing, stores refrigeration proportional to the

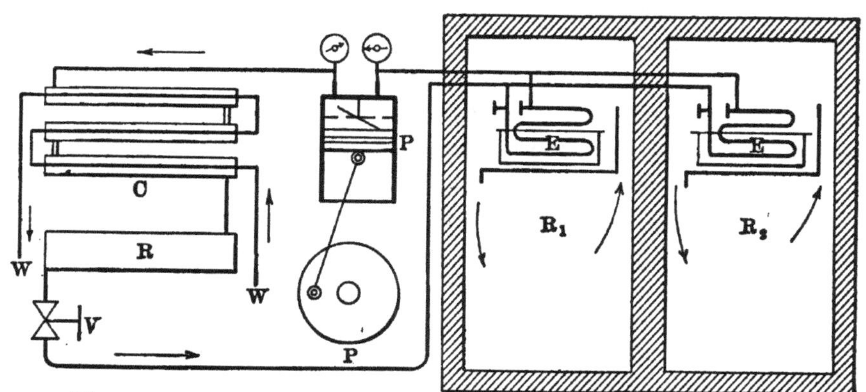

Fig. 9.—Congealing Tank—Compression Refrigerating System

number of pounds of brine frozen multiplied by the latent heat of fusion of the brine ice plus the cold required to chill the brine down to the freezing point:

$$B.t.u. = lb.\ ice \times \{latent\ heat + [specific\ heat\ of\ brine \times (t - t_1)]\}.$$

This expression does not provide for the small additional amount of refrigeration required to cool the brine-ice below its freezing point. When the temperature of the ice is below the freezing point the additional refrigeration will be found from the expression

$$B.t.u. = lb.\ ice \times specific\ heat\ of\ the\ ice \times (t_1 - t_2).$$

In the above expressions t is the temperature of the brine before cooling, t_1 the temperature at which the brine freezes and t_2 the temperature to which the ice is cooled after freezing.

In order that the rooms may be cooled more quickly when the refrigerating plant resumes operation after several hours of inaction, it is often desirable to install from one-third to two-thirds of the total expansion piping outside the tanks in direct contact with the atmosphere of the rooms, the remaining two-thirds or one-third being submerged in the weak brine of the congealing tanks.

It is evident from the foregoing that so far as decreased efficiency entailed by the double-heat interchange of the brine-circulating system is concerned, the congealing-tank system is only a compromise. While a part of the piping cools the air by direct radiation, the greater part must transmit cold first to the surrounding brine, or ice, as the case may be, which medium in turn must transmit it on through the walls of the tanks to the air; and a sufficiently low ammonia evaporation or "back pressure" with its entailed loss of efficiency, must be maintained in both the submerged and the exposed coils, to carry out the double-heat interchange in the latter. The fact that the refrigeration stored up during the hours of operation of the plant for use during the hours of rest is in the form of a coating of ice on the coils, instead of brine kept in constant circulation, further increases the necessity for a lower back pressure; first, because ice is a poorer conductor of heat than brine; second, because the congealing-tank surface, as well as that of the expansion pipes, is often insulated with ice; and third, because the stagnant brine, or, what is even worse, ice in the case of the congealing-tank system, absorbs and gives up heat less readily than the moving brine.

As it is mechanically impracticable to make the thin, light congealing tanks indefinitely continuous, as is done in the case of ammonia or brine-cooling coils, the ammonia-expansion coils installed in such tanks must be made shorter. This necessitates the use of an increased number of return bends or fittings for the installation of the system as a whole and accordingly tends to increase the initial cost of the expansion piping for both material and labor.

Electrically Driven Plants

While the use of electric power undoubtedly reduces the duties of the attendant, safety and the necessity of adjusting expansion valves by hand require more or less constant attention, and the slight additional attention required by the steam or combustion engine is usually too slight to make up for the usual comparatively high cost of electric power.

The application of reliable safety devices, however, which protect the plant in case of abnormal pressure resulting from failure of water supply or the accidental closing of the wrong valves, together with a reliable automatic expansion valve, eliminates these two most important duties of the attendant. When the compressors are of the inclosed-crank self-oiling type, and electric motors are employed, very little attention need be paid to lubrication, and the item of attendance becomes very small indeed.

Semi-Automatic Systems

The semi-automatic system usually employed to meet such conditions is illustrated diagrammatically in Fig. 10. It will be

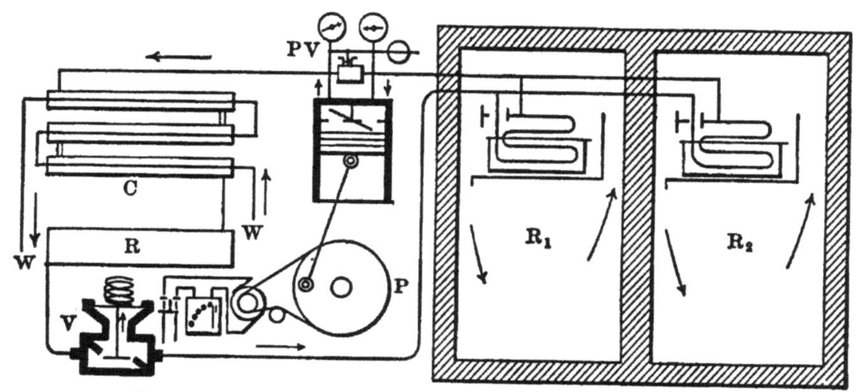

Fig. 10.—Semi-Automatic Congealing Tank System

noted that this system is essentially the same as the congealing-tank system illustrated in Fig. 9, except for the addition of an automatic expansion valve *V* and some suitable safety device *P V* for preventing the occurrence of abnormal pressures. In the case of other than electrical power, this may be a simple mechanical device, such as a spring or a weight-loaded safety valve between the high-pressure and the low-pressure sides of the system. In the case of electric power a device actuated by pressure, such as is shown diagrammatically in Fig. 11, may be employed. Such a device consists essentially of either a Bourdon tube or a diaphragm *D* arranged to move a lever arm *A* in such a way as either to disengage a latch holding a spring-opening electric switch closed or, where there is a no-voltage release coil in the motor-controlling circuit, the device may be arranged as in *P D*, Fig. 12. Here the diaphragm *D* bowing outward under abnormal pressure *P* is employed simply to make an electric contact *C* for short-circuiting a no-voltage coil which, being thereby deënergized, allows a spring to open the motor circuit and interrupt the operation of the refrigerating system, just as would occur in case of failure of line voltage. It is almost unnecessary to add that a thermostat *T* may be similarly employed to stop the refrigerating machine when the desired temperature has been produced in the cold-storage compartments.

In the case of a spring-opening switch an outside source of power, such as a storage battery, would have to be used in connection with the thermostat to supply sufficient power to operate the latch. Instead of a laminated-blade form of thermostat controlling a battery circuit, a thermostatic device *T D*, Fig. 12, worked

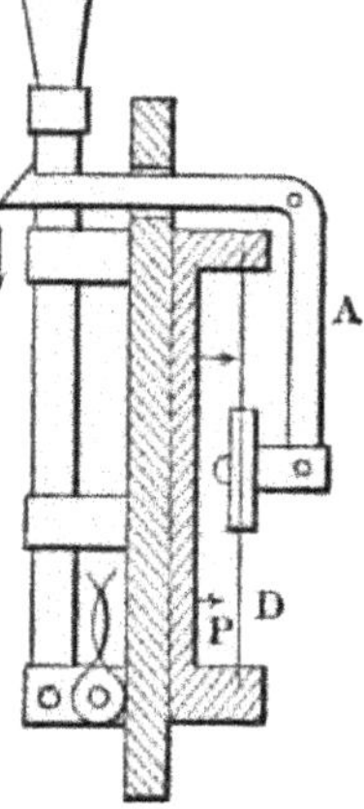

Fig. 11.—Pressure Actuated Circuit Breaking Device.

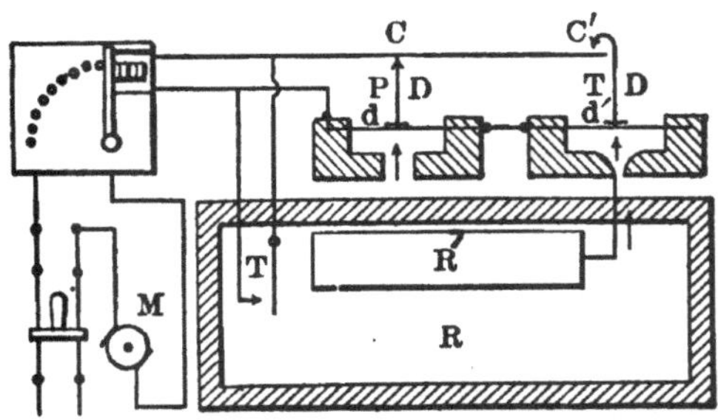

Fig. 12.—Electrically Actuated Circuit Breaking Device

by the expansion force of a gas under pressure may be employed. In this case a small amount of some volatile liquid, such as anhydrous ammonia, is placed in a small closed receiver *R'* and connected to the thermostatic device *T D* by a small tube. Any decreased temperature in the cold-storage compartment *R* causes a corresponding decrease in pressure in the receiver *R'* and beneath the diaphragm *d'*, which allows it to return to its normal flat position where it effects the making of an electrical contact at *C'* which stops the machine.

Completely Automatic Systems

Still more elaborate systems have been devised to the end of eliminating all attendance. "Completely automatic" systems divide themselves into two classes according to the two possible cycles of cause and effect on which the systems may be operated. In the first system, illustrated diagrammatically in Fig. 13, variations in temperature in the cold-storage room *R* effect the starting and stopping of the refrigerating machine, which through the resulting variations in back pressure effects the regulation of the flow of refrigerant to the expansion coils and through the resulting variations in head pressure effects the regulation of the flow of water on the condenser coils.

In the second system, illustrated diagrammatically in Fig. 14, variations in temperature in the cold-storage room *R*, instead of effecting the starting and stopping of the refrigerating machine, and indirectly the regulation of the flow of the refrigerating medium, in this case directly controls the flow of the refrigerant through a thermostatic-expansion valve *V*, which in turn controls the starting and stopping of the machine through the resulting pressures. The cycle of operation of each system is followed out in detail below.

Thermostat Control

The thermostatically controlled system shown in Fig. 13 employs a laminated-blade or other reliable type of thermostat for making and breaking two electric circuits which, without going into full detail regarding the complete circuit, stops and starts

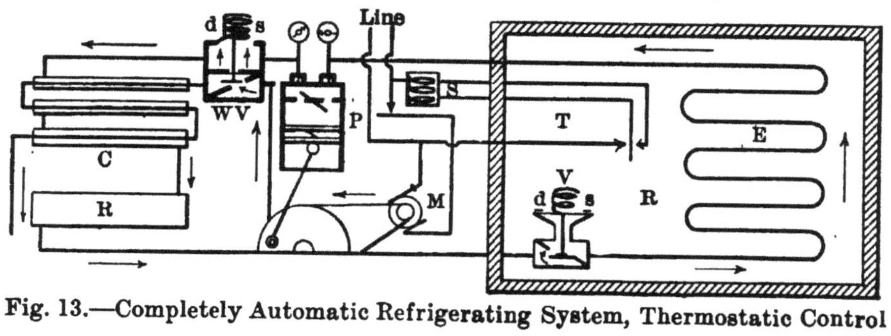

Fig. 13.—Completely Automatic Refrigerating System, Thermostatic Control

the machine through the operation of an appropriate automatic motor-controlling panel represented diagrammatically by the electrical solenoid *S*, which, when energized, raises the motor-controller arm slowly over the rheostat segments. The deënergizing of this solenoid through short-circuiting, when the thermostat makes the circuit through the stopping contact, allows the rheostat arm to drop, open circuiting and stopping the motor. The operation of the machine is to draw the refrigerant vapor from the expansion coils *E* and discharge it into the condenser *C*, which operation reduces the pressure in the former and increases the pressure in the latter coils. The reduction in pressure in the expansion or low-pressure side of the system allows a properly adjusted spring *s* or weight on the expansion valve *V* to overcome the back pressure of the gas exerted beneath the diaphragm *d*, which forces the attached valve stem down, opens the valve and allows the refrigerant to flow from the liquid receiver *R* to the expansion coils. In a similar manner the increased head or condenser pressure operating on a water controlling valve *W V*, which instead of closing with increasing pressure, opens and admits ccoling water to the condenser *C*.

The evaporation of the liquid refrigerant admitted to the expansion coils by the expansion valve produces an increase in pressure which overcomes the pressure of the spring above the diaphragm and closes the valve until the pressure is so reduced by the drawing away of the vapors by the compressor that the spring again overcomes the pressure of the gas below the diaphragm, when the valve opens and more liquid is allowed to pass. As a matter of fact, there is no appreciable variation in back pressure and accordingly no intermittence in the feed of the liquid. The valve acts simply as a pressure-reducing valve, maintaining that back pressure for which it is adjusted until the machine stops, when the evaporation of the residual liquid in the expansion coils produces an abnormal rise in back pressure which overcomes the tension of the spring and tightly closes the valve.

When the machine is stopped, and no more hot high-pressure gas is discharged into the condenser, the cooling water soon reduces the temperature and corresponding pressure in the high-pressure side of the system, which allows the spring in the water valve *W V* to overcome this reduced pressure and close the valve, interrupting the flow of water to the condenser.

Expansion Valve Control

In the second method of control, illustrated in Fig. 14, an increase in temperature causes the thermostatic fluid in the ther-

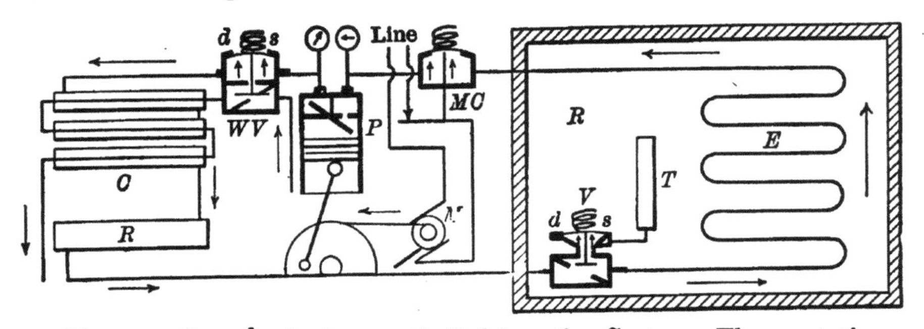

Fig. 14.—Completely Automatic Refrigerating System—Thermostatic Expansion Valve Control

mostatic tube *T* to expand and exert an increased pressure under the diaphragm *d* of the thermostatic expansion valve *V*, overcoming the adjusting spring *s*, opening the valve and admitting liquid refrigerant to the expansion coils. This increase in expansion-coil pressure, caused by the introduction of the liquid, overcomes the pressure of the spring above the diaphragm of the pressure-actuated, motor-controlling device *M C* making the electric circuit to the motor and starting the machine. The resulting decrease in back pressure allows the spring of the expansion valve to arrest itself, forcing the diaphragm down and throttling the incoming liquid to the requirements of the compartment. When the temperature has been sufficiently reduced, the correspondingly reduced pressure in the thermostatic tube *T* allows the spring *s* to close the thermostatic expansion valve *V* and the reducing pressure in the expansion coils is now overcome by the spring in the motor-controlling device *M C*, which, operating in the opposite direction due to the decreased pressure, stops the motor.

The automatic water valve shown in Fig. 14 operates on the same principle as that described in Fig. 13. In the semi-automatic systems illustrated diagrammatically in Figs. 10 and 12, as well as in the completely automatic systems, Figs. 13 and 14, electric power is employed.

Automatic Control of Feeds in Parallel

The automatic systems above described have quite serious limitations because of the arrangement of all the expansion coils in series, a condition which not only limits the ability of the system to accurately maintain the various temperatures often required, but also imposes unusual friction to the passage of the refrigerating fluid. This second condition makes it necessary to operate the compressor at a lower back pressure than would otherwise be necessary, in order to produce the desired temperatures at the expansion valve end of the expansion coils.

While little has as yet been accomplished commercially, a system has been devised and patented by Stephen C. Wolcott for the automatic control of systems in which the several cold storage compartments are each provided with expansion coils installed in parallel and each fed by an independent liquid pressure reducing valve, controlled by the temperatures in the respective compartments.

In the Wolcott system the motor circuit is so arranged that the compressor will continue to operate so long as the temperature in any one of the compartments is sufficiently high to keep its controlling thermostat on the operating contact. As the temperatures of the several compartments are successively reduced, the thermostats make low-temperature contacts, which, by means of springs, weights, or other outside agencies, effect the closing of stop valves placed in the return ends of the expansion coils. As each low-temperature or motor-stopping contact is made by the thermostats, the corresponding high-temperature or motor-starting contact, arranged in parallel for the control of the main motor circuit, is broken. When the last compartment has been sufficiently cooled and the last of the parallel controlling circuits broken, the last stop valve is closed and the machine is made to suspend operation until a sufficient rise in temperature in some one of the compartments causes the controlling thermostat of that compartment to open the stop valve of the expansion coil in question, and at the same time to actuate the controlling mechanism to start the motor.

The reduction in back pressure in each open coil, due to the operation of the compressor, causes the liquid pressure reducing valve, controlling the feed of that coil, to admit refrigerant as required until the rise in back pressure, due to the evaporation of the refrigerant remaining in the coil when the thermostat effects the closing of the stop valve on the end of the coil, again closes it.

Among the numerous types of small refrigerating machines designed for capacities suitable for household and other purposes where only a few hundred pounds of cooling effect are required per twenty-four hours, the following are the most interesting from the viewpoint of mechanical construction, one being an absorption and the other a compression type of machine.

A Small Capacity Absorption Machine

The absorption machine described below, the invention of Dr. J. W. Morris, is of particular interest in that it employs water as a working medium, instead of ammonia, and caustic potash or potassium hydrate as an absorbing medium instead of water. In number and functions the several parts of the Morris machine are similar to those of the usual absorption machine employing ammonia as a working medium. Their construction, however, is very different, since to employ water as a medium to extract heat at the freezing point or below requires a vacuum of 29.75 inches and over. Dr. Morris's experimental machine employed glass tubes in the place of iron pipes, the joints being made by means of short pieces of rubber tubing wound with thread and painted to ensure air tightness. The machine is started under high vacuum, which is further increased by the high affinity which the absorbing medium has for water. Under the high vacuum maintained in the refrigerator, the working medium (water) boils at temperatures of 32° and slightly lower, depending on the pressure, sodium chloride or salt being dissolved in the water to keep it from freezing. The vapor or "steam" passes from the refrigerator to the absorber—where it unites with the concentrated potassium hydrate (strong liquor) to form dilute potassium hydrate (weak liquor), which is pumped over to the generator (still). Here the water vapors are driven off to the condenser, after which the resulting strong potassium hydrate is allowed to flow back to the absorber, where it absorbs more aqueous vapor coming from the refrigerator by way of the condenser.

The still is heated by a small gas flame under automatic

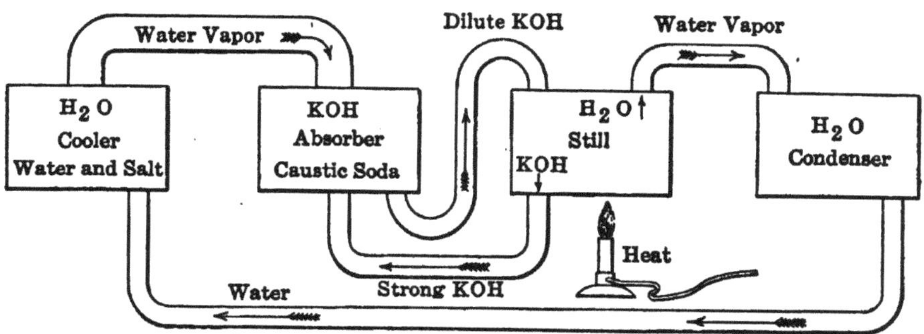

Fig. 15.—Morris Absorption Machine

control. As in the usual type of absorption machine, the weak liquor flows from the generator to the absorber because of the higher pressure in the former, the only place where power is necessary being where the strong liquor is passed from the absorber back to the generator. Here a "potassium hydrate" pump, paralleling in action the ammonia pump of the usual system, is required. In the Morris machine this pump consists of a little hydraulic ram actuated by the pressure of the cooling water on its way to the condenser.

A Small Capacity Compression Machine

The compression type of machine designed to meet the requirements of the small consumer of refrigeration seeks to overcome one of the often prohibitive features of the usual types when adapted to household use, viz., the escaping of the working fluid through packed glands and gasket joints, by enclosing the whole mechanism in a hermetically sealed case. The machine in question, designed and patented in 1895 by a Frenchman named Marcel Audiffren, consists essentially of two small gas compressors suspended on a shaft which runs longitudinally through a dumb-bell-shaped enclosing casing.. This shaft is provided with cranks in line with the compressor cylinders and the whole is so arranged that when the casing and shaft are revolved and the compressor remains stationary, a reciprocating motion is given to the compressor piston as in the usual type. The casing, mounted on suitable bearings, with its axis horizontal, is half submerged in a tank so arranged that one end of the dumb-bell-shaped casing revolves in a liquid on one side of a partition through this tank, while the other part containing the compressor revolves in another liquid on the other side of the partition. Two ducts are provided through the shaft connecting the two parts of the casing, one of which forms a suction line to the compressor and the other an expansion or liquid line to the other chamber.

When charged with a refrigerating liquid, the tank surrounding

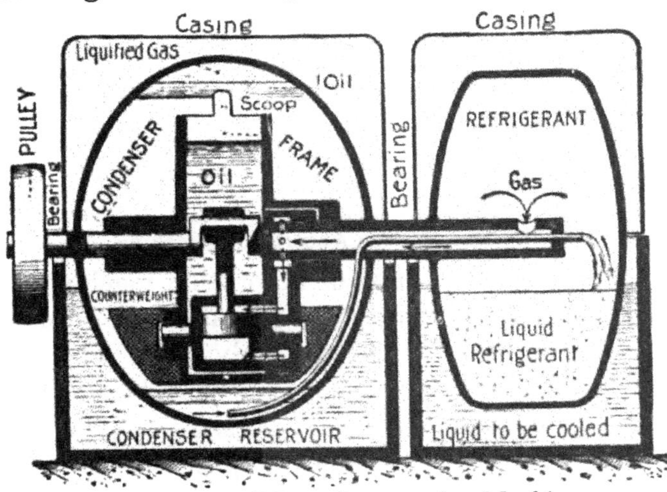

Fig. 16.—Audiffren Compression Machine

the casing containing the compressor is filled with cooling water, that surrounding the other casing with brine, and the machine placed in operation. The action of the compressor is to draw the vaporized gaseous refrigerant from the expansion chamber, compress it, and discharge it into its condensing chamber, where, being surrounded by cooling water, it again liquefies under the higher pressure. The liquid refrigerant, thrown to that part of the casing most remote from the axis of rotation, enters the liquid line and is conducted to the other compartment, where evaporating at a lower pressure, it refrigerates the brine surrounding the refrigerator chamber, and finally returns as a gas to the compressor to be again liquefied by the cooling water surrounding the condenser chamber.

The Selection of a Refrigerating System

While local conditions must be the deciding factors in every case, congealing-tank systems of small capacity usually compare quite favorably with the brine-circulation systems of similar capacities; but it must be remembered that both systems have the disadvantage of being inherently inefficient because of the necessity of operating under lower back pressures in order to produce the correspondingly lower temperatures required to effect the double-heat transfer. At the lower pressures fewer pounds of the refrigerating medium are passed through the compressor per revolution and a greater number of pounds are necessary to produce a given refrigerating effect, each of which conditions makes it necessary to speed up the compressor in order to produce the same amount of cooling duty.

By the direct expansion of the refrigerant, cold is produced just where it is required and more nearly at the temperatures required. And incidentally, aside from the saving in power required to operate the compressor and to pump the brine, a saving is made in investment for a suitable room and for foundations of brine tanks and pumps, as well as for the tanks and pumps themselves and the necessary piping and insulation, etc. Thermal losses due to radiation through the insulation of these members and through that due to the heating effect of pumping the brine are entirely eliminated.

Small systems may be satisfactorily operated semi-automatically at the expense of somewhat more attendance, and usually effect a considerable saving in cost of power by the substitution of some efficient type of combustion engine. The advisability of such a substitution at the present day of an unquestionably reliable combustion engine depends almost entirely on the relative local cost of attendance, and the cost of power developed by electricity, illuminating gas, gasolene, kerosene, fuel oil or producer gas.

The type of system best adapted to a given set of requirements can be determined only after carefully considering the relative importance of such factors as cost of power, cost of attendance, allowable temperature variation, probable earning power of plant and availability of capital.

CHAPTER IV

A. The Compression System

While it is impossible to show in a single illustration all the details entering into the mechanical construction of a complete refrigerating system, the diagrammatic representation, Fig. 17, is more complete than the elementary diagrams already shown, and

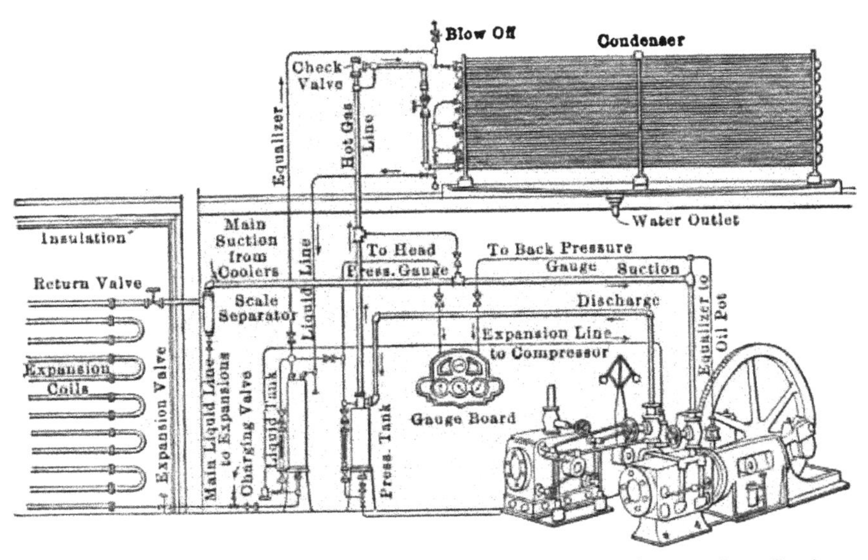

Fig. 17.—General Arrangement of Compression Refrigerating System

will suffice for the explanation of the cycle of operation of the compression system.

Expansion Side

The cycle begins with liquid anhydrous ammonia conducted to the expansion coils through a liquid line and regulated by appropriate expansion valves. The source of the liquid ammonia may be either a liquid receiver, which, as already shown, is one of the essential members of the system in practical operation; or it may be an ammonia shipping-drum from which ammonia is introduced into the system by the process of charging. In either case the first function of the refrigerant is to enter the expansion coils in the compartment to be refrigerated and there to evaporate (boiling at a temperature dependent on the suction or "back pressure" within the coils) and absorb heat from the surrounding objects. If the back pressure be 46 pounds gauge, or less, the temperature of the boiling liquid will be 32° Fahrenheit, the freezing point of water, or below, and the pipes containing the refrigerating medium at these temperatures will soon be covered with a coating of frost. If, on the other hand, the back pressure in the coils is above 46 pounds gauge, the temperature of the boiling liquid will be above 32° Fahrenheit, and no frost will be formed. Refrigeration will be produced whenever the temperature of the expansion coils is lower than that of the air in the cooler, regardless of whether frost is formed or not and, according to the amount of atmospheric humidity, uncongealed moisture may or may not be precipitated on the pipes.

Having vaporized in the expansion coils, the ammonia vapor enters the return header which conveys it back to the suction side of the compressor. This return line is usually fitted with a "scale trap" constructed similarly to a simple steam separator. The function of this trap is to prevent any scale from the inside of the pipes, or other foreign substance, from entering and damaging the compressor. From the scale trap the gas passes through the suction valves into the compressor.

Compression Side

On leaving the compressor the hot, high-pressure ammonia gas passes through the discharge valves and out into the two legs of the main discharge pipe, which come together in a "T" just below the large hand-operated discharge valve at the left-hand side of the compressor cylinder. Leaving the discharge valve, the gas passes into the side of the pressure-tank head, and is given a spiral motion as it descends into the tank. The centrifugal force produced by this spiral motion is intended to aid in precipitating any entrained oil which the gas may hold in suspension against the outside of the tank. The gas then passes up through the hot gas line, through a "check valve" (sometimes omitted) located a little above the level of the top of the condenser, and down into the header connecting the bottom pipes of the several stands of condensers. The loop is placed in the hot-gas line to prevent condensing liquid from running back down into the pressure tank when the compressor is shut down.

The gas entering the bottom pipe of the condenser passes up through successive pipes while the water distributed over the top pipe trickles down, producing the desirable countercurrent cooling effect, in which the hottest water encounters the hottest gas at the bottom and the coolest water the coolest gas at the top of the condenser.

As fast as the ammonia is condensed in the pipes of the condenser it is conducted away through smaller liquid-drip pipes connected with a common liquid header. The outlet from this header also rises to form a short "goose-neck" which is intended to keep the header always full of liquid and prevents gas from being drawn down to the tank through the liquid line. In trying to ascend through the column of descending liquid in a small liquid line, bubbles of gas offer no inconsiderable resistance to the passage of the liquid. The obvious remedy for this difficulty is the installation of lines of liberal diameter.

If the gas is carried down into the liquid tank it can escape by going up the equalizer line into the top of the condenser. The ammonia previously liquefied in the condenser under a pressure of from 135 to 200 pounds, according to the temperature of the cooling water, is conveyed first to the receiver. This consists of an appropriate cylindrical containing vessel which acts as a storage tank in which the liquefied ammonia is collected and from which it passes as required into the expansion coils. The flow of this high-pressure liquid into the expansion coils is regulated by expansion valves, which are virtually nothing more than convenient mechanical devices for accurately varying the opening through which the liquid ammonia must pass on its way to the expansion coils.

Refrigeration Available in Expansion

As stated, the word "expansion" has been erroneously applied to these valves and coils, because of the idea, also erroneous, that the liquid ammonia vaporizes or expands immediately when the pressure is relieved as it passes the regulating valve and enters the cooling coils. As a matter of fact, before it is possible for a pound of ammonia to change from the liquid to the gaseous state, it must be supplied with about 573 B.t.u. of heat.*

In practice, not all of the heat-absorbing capacity, or negative heat, of a pound of anhydrous ammonia available at 0° Fahrenheit can be utilized for useful cooling work, on account of the cooling work which must first be expended on the ammonia

*Evaporation assumed to be under atmospheric pressure.

itself in order to reduce its temperature from that of the condenser to that of the cooler. This may be illustrated by a similar process in which water is the medium. The amount of heat that must be abstracted from one pound of water at 32° Fahrenheit in order to freeze it is 144 B.t.u. On this basis, a ton of ice would represent 288,000 B.t.u. of negative heat. In practice, the expenditure of this amount of cooling will not freeze a ton of water, because it must first be reduced from its natural temperature, or, in crystal-ice systems, from the temperature of the distilled water tank to 32° Fahrenheit. This involves a further expenditure of one negative B.t.u. per pound per degree cooled.

If the 573 B.t.u. were absorbed at the expansion valve, which its immediate vaporization assumes, there would be no further heat-absorbing capacity in the ammonia, and its introduction into the expansion coils would be useless.

Besides the principal pipe circuit just described, the compressor is provided with a set of bypass connections for reversing its operation so as to draw the ammonia from the condenser and discharge it into the expansion coils, as well as other so called "pump-out" lines through which ammonia may be pumped out of other parts of the system in case it becomes necessary, as when making repairs.

Direct Expansion Cylinder Cooling

In Fig. 17 is shown a small liquid line running from the liquid tank to the compressor cylinder. This line is provided with an expansion valve through which ammonia may be admitted to the compressor, to prevent abnormal heating of the piston and packing when starting up, or at any other time when the ammonia returning to the compressor is not sufficiently cold to insure satisfactory operating of the compressor. Another small pipe line connects the lubricating system on the compressor with the pressure tank. Through this line oil passing over with the discharged ammonia gas and separated out in the pressure tank may be blown back into the lubricating system. On entering this line the oil passes first through a small strainer which intercepts any scale or foreign substances that might otherwise return to the compressor and cut the valves or cylinder walls.

Details of construction of the various parts of a compression system are too numerous to warrant an attempt to fully describe them here. Since the ammonia compressor is so important a member of the refrigerating system, however, a brief description setting forth its several characteristics will be in order.

Types of Ammonia Compressors

Ammonia compressors are divided into two principal classes, double-acting and single-acting. The former type is most commonly horizontal, although frequently of vertical construction. The single-acting type is almost exclusively a vertical machine. Each type has its own followers among builders, and under certain conditions possesses some advantages over the other. While there is much variation in details of design among the various builders,

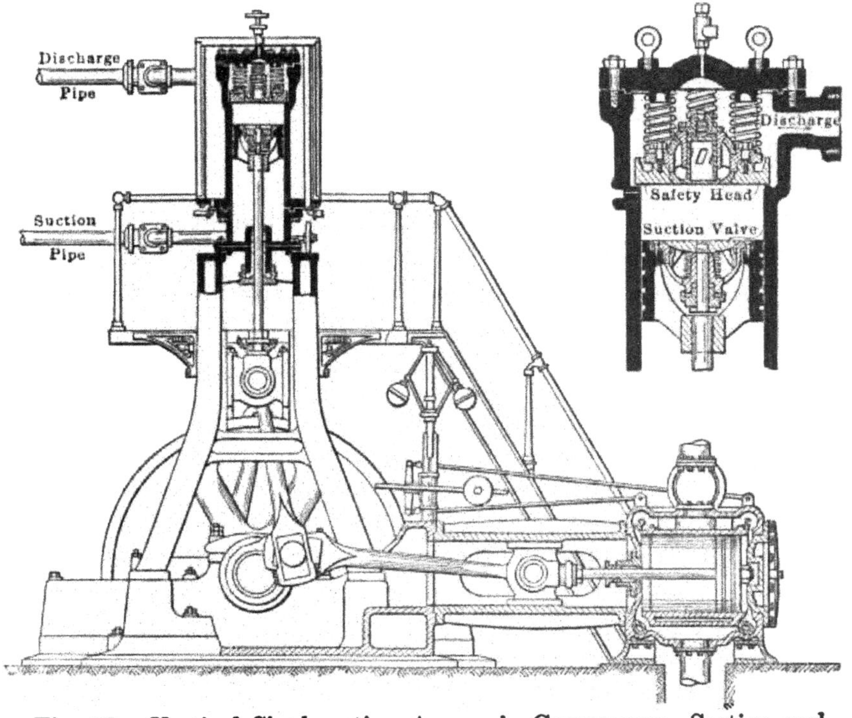

Fig. 18.—Vertical Single-acting Ammonia Compressor—Section and Typical Elevation.

the accompanying illustrations, Figs. 18 and 19, are the most characteristic of the general types. Fig. 20 shows a modification of the vertical single-acting machine which may be said to be typical of the compressor of small capacity.

Vertical Single-acting Compressors

The accompanying illustration, Fig. 18, giving a lateral elevation in section of a characteristic vertical single-acting ammonia compressor, shows the relative arrangement of compressor and engine cylinders as well as the principal details of design. In this type of compressor the vaporized refrigerant enters the compressor near the bottom, passes up through the suction valve, located in the compressor piston, during the downstroke, and is compressed and discharged through the discharge valve, located in the safety head, during the upstroke of the piston. The compressor may be water jacketed or not. Popular preference, however, is in favor of the water jacket, and most machines are so built.

Vertical compressors possess the advantage of requiring less floor space than horizontal machines and the disadvantage of being less accessible for repairs. The inaccessibility of suction valves located in compressor pistons is offset by the unmistakable advantage offered by this type of machine in that these suction valves can be made of generous area, and the inertia of the valve tends to hasten its closing promptly as the piston reverses its direction of travel at the lower end of its stroke. This largely prevents gas from escaping from the cylinder during the time required for the acting of the ordinary stationary valves unless they be heavily spring loaded, a condition which tends to prevent the back pressure in the cylinder from quite reaching the height of that in the suction line from the coolers. Inertia also tends to open the valve immediately as soon as the piston begins its downward stroke, giving full opportunity for the cylinder to fill. The spring below the suction valve should be of such strength as to almost balance the weight of the valve, so that the inertia of the valve may act promptly at each end of the stroke.

Vertical single-acting compressors are usually provided with a "safety head" which is normally held securely to its seat by strong springs, but which, in the event of abnormal quantities of liquid ammonia or broken parts entering the cylinder above the piston, is pushed back, compressing the springs and thereby saving the machine from the strains that would otherwise occur. One of the principal advantages claimed by the advocates of the single-acting compressor is that the use of the safety head allows the compressor pistons to be operated with less clearance than would be practicable in the case of double-acting machines, a condition which insures a more complete expulsion of the gas.

The Horizontal Double-acting Machine

Fig. 19 represents a horizontal half section of a characteristic horizontal double-acting ammonia compressor. The right-hand portion of the cut shows the exterior of the head end of the compressor-cylinder valve housings, suction and discharge connections and valves. The remaining portion of the cut shows the details of construction of the compressor cylinder, water jacket, piston,

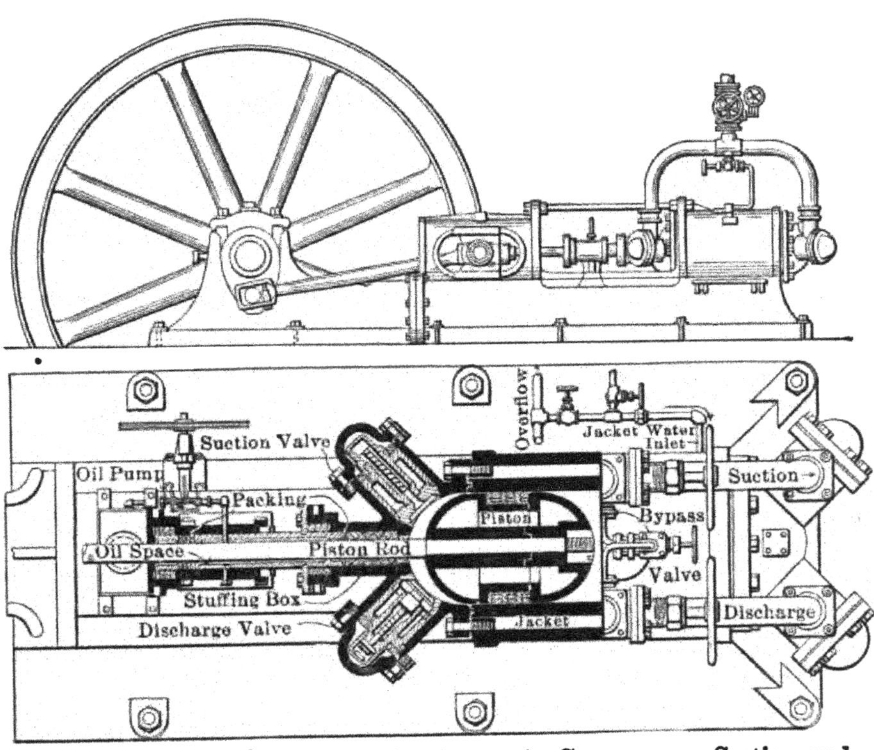

Fig 19.—Horizontal Double-acting Ammonia Compressor—Section and Typical Elevation.

suction and discharge valves, double stuffing box and means of lubricating the piston rod. The outer wall of the water jacket is formed by the main frame casting, which is bored and fitted with a working cylinder liner, consisting of a straight sleeve forced into place by hydraulic pressure and bored to the required size. The valves in this type of compressor are arranged radially to the hemispherical cylinder heads. The piston rod is provided with a primary stuffing box where it enters the compression cylinder. The packing in this box is tightened by a primary packing nut which carries a long sleeve, the other end of which is provided with a secondary stuffing box and packing nut. The main stuffing box, containing the bulk of the packing, withstands the high pressure of the ammonia in the compressor cylinder. The small stuffing box at the end of the sleeve is provided with sufficient packing to

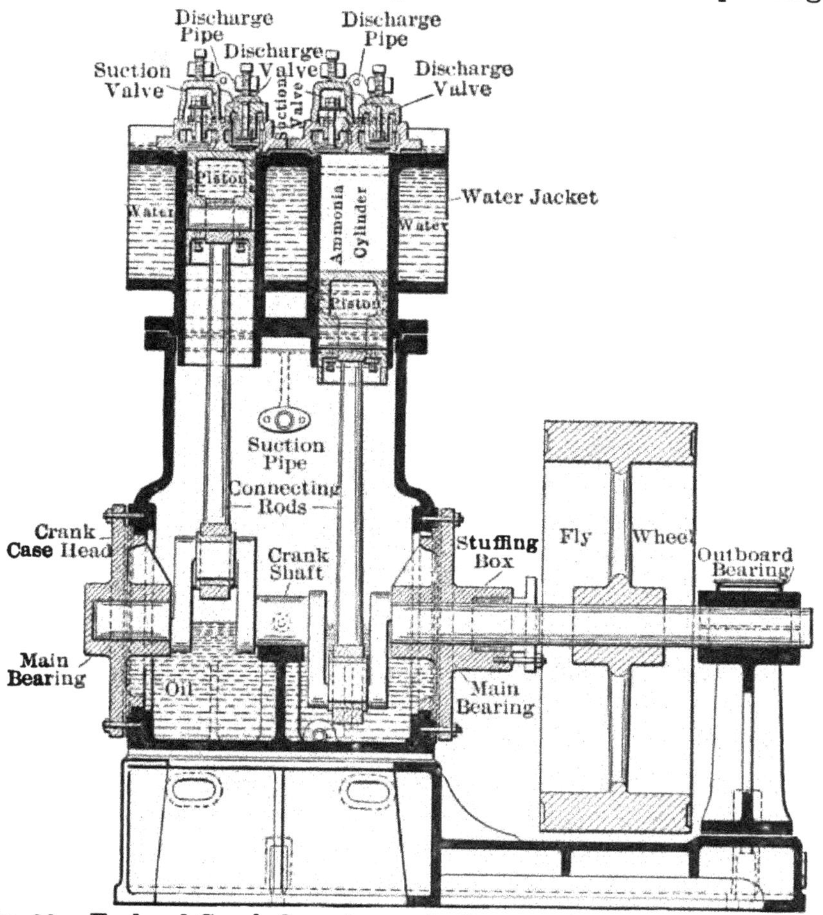

Fig. 20.—Enclosed Crank Case Ammonia Compressor—Elevation in Section

withstand the pressure of the oil circulated by the oil pump through the hollow sleeve surrounding the piston rod, in order to insure constant lubrication of and to maintain an oil seal on the main stuffing box.

The general appearance of the horizontal double-acting compressor just described is shown in the longitudinal elevation, just above the sectional view of the compressor cylinder.

Inclosed Crank-case Compressors

In addition to the two principal types of compressors previously described, the inclosed-crank type is deserving of mention because of the great number of such machines of small capacity now being installed. Details of design of this type of compressor are even more varied than those of the machines already described, and it is difficult to point out a single design that can be said to be more characteristic of the type than another.

In the illustration, Fig. 20, the refrigerant vapor enters the compressor cylinders through suction valves located in the cylinder head. Valves so located cannot be made of so liberal dimensions as those located in the compressor piston, and the assistance which inertia offers in the way of opening and closing suction valves located in the piston cannot be realized. To offset this disadvantage, oil from the crank case is much less likely to be carried over into the condenser and low-pressure side of the system.

Machines of the inclosed type are especially adapted to use where little, or only inefficient, attendance is available. Less attention is possible in this type of machine, principally because of the design of the stuffing box and the main-bearing lubrication. The crank case being filled with oil to the center of the crank shaft, and the outboard bearing being usually ring oiling or provided with a compression grease cup, little attention to lubrication is necessary. There are no reciprocating piston rods to pack, the only stuffing box required being on the crank shaft, where it is always well lubricated and not subject to such extremes of temperature as are the pistons in other types of machines. This type of compressor is peculiarly well adapted for use in connection with automatic systems.

B. The Absorption Refrigerating System

It has already been pointed out under the subject of "The Development of Mechanical Refrigeration" that, while Carré invented the absorption machine, the way was paved by the earlier experiments of Faraday, who discovered that silver chloride possessed the property of absorbing ammonia. Faraday is said to have experimented with silver chloride saturated with ammonia in a closed glass tube, one end of which was immersed in a freezing mixture of ice and salt, while the other end was heated. The ammonia gas driven off from the silver chloride in the hot end of the tube was liquefied in the cold end of the tube under the pressure generated by the heat. See Fig. 21.

Faraday discovered that if he reversed the tube so that the

end containing the silver chloride from which the ammonia had been driven off was immersed in the freezing mixture, the liquefied

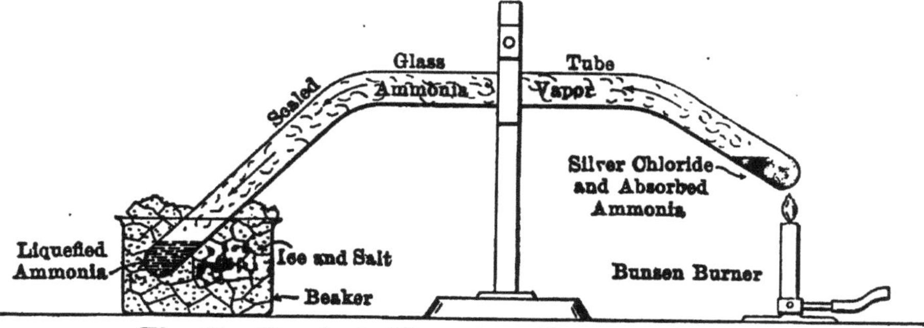

Fig. 21.—Faraday's Elementary Absorption Machine

ammonia in the other end of the tube would boil, producing a remarkably low temperature. The cold vaporized ammonia was absorbed by the silver chloride, so that if the tube were occasionally reversed the device would be made to traverse the cycle of an elementary, intermittent absorption machine.*

A modern commercial absorption machine consists primarily of five parts, three of which are also present in the compression machine. This is shown in Fig. 22, which is a diagrammatic representation of an elementary absorption and compression machine having their condensers and complete expansion sides in common.

In the compression system it has been shown how the low-pressure cold gas returning from the expansion coils enters the suction end of the compressor cylinder, and how, after it has been transferred to the compression end of the cylinder, it is compressed and passed over into the condenser. Reference to the figure will show that in the absorption machine the compressor cylinder is replaced by an absorber, a liquid pump and a generator. In the absorption system the gas, returning from the expansion coils, enters the absorber (corresponding to the suction end of the compressor), is transferred to the generator (corresponding to the discharge end of the compressor) by a pump, through the valves of which it passes just as it flows through the valves of the compressor piston.

In the absorption plant the ammonia (liquor) pump can be made much smaller than the compressor gas pump used in the compression system, because the actual work of compressing the

* This experiment of Faraday's is described more in detail by Mr. Gardner T. Voorhees in "Ice and Refrigeration," October, 1908.

ammonia gas to the point at which it can be liquefied by the cooling water in the condenser is performed by the direct heat of steam rather than by the heat generated by the expenditure of energy behind the compressor piston. To facilitate the use of steam for this purpose, the cold ammonia gas returning from the expansion coils is absorbed or dissolved in water in the absorber, and the resulting strong aqua ammonia, known as "strong liquor," is pumped into the generator or ammonia boiler, where it is heated by a series of steam coils which drive off the gaseous ammonia at a high pressure, just as water vapor or steam is driven off under high pressure in a steam boiler.

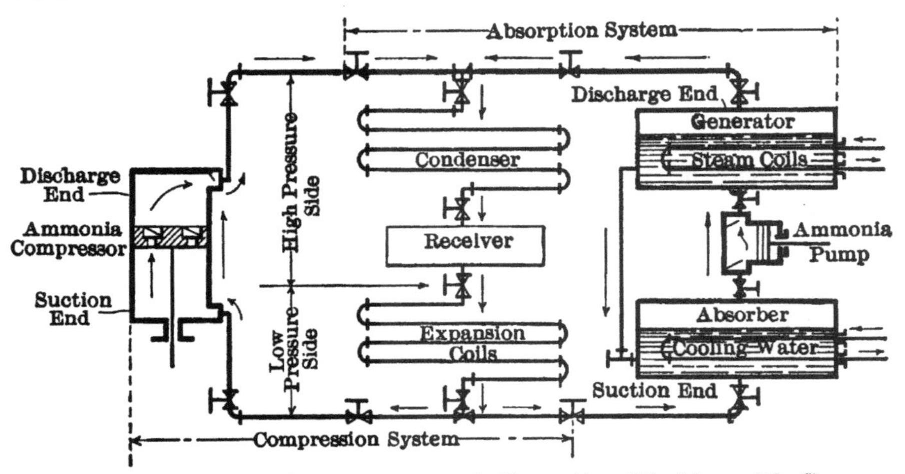

Fig. 22.—Elementary Compression and Absorption Machine with Common Condensing and Expansion Units

The high-pressure ammonia gas driven off in the generator of an absorption plant, like that discharged from the compressor of a compression plant, is conducted to the condenser, at which point that part of the refrigerating cycle common to both systems begins.

The driving off of the ammonia vapor in the generator changes the strong liquor to "weak liquor," which, being under a higher pressure than the liquid in the absorber, readily flows back to the absorber. To adapt the elementary absorption machine just described to the requirements of economical commercial apparatus, a number of refinements must be added. Fig. 23 is a diagrammatic representation of a modern absorption machine of well-known make.

On account of the similarity of the expansion side of the absorption system to that of the compression system already de-

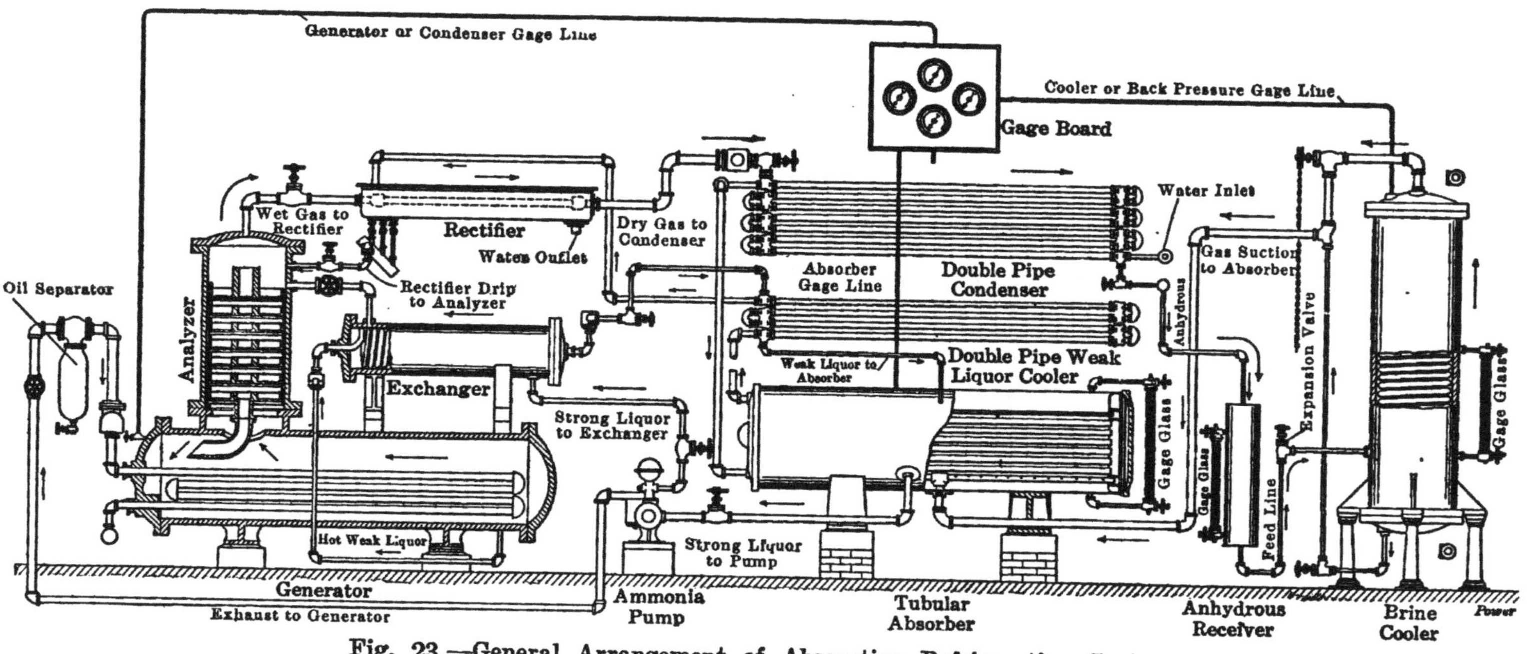

Fig. 23.—General Arrangement of Absorption Refrigerating System

scribed in detail, it will be necessary to trace the working medium through only that part of the cycle between the expansion coils and the condenser.

Cooler

In the type of machine illustrated in Fig. 23, the expansion side consists of a brine cooler of the vertical cylindrical or shell type, such as is most commonly used in connection with absorption machines, but to some considerable extent with compression machines as well. The brine, usually calcium chloride, is circulated through nests of spiral coils within the cylindrical shell of the cooler, and the anhydrous ammonia is expanded into the space between the coils and the shell. The amount of liquid present is readily observed by means of a gauge glass. The cold ammonia vapor leaving the top of the cooler passes through the gas-suction line to the absorber, where it is joined by the weak liquor which has just undergone cooling in the "double-pipe" weak-liquor cooler. Since the absorption of the ammonia gas into the weak liquor takes place with the evolution of a considerable amount of heat, further means for cooling the liquor must also be provided in the absorber.

Absorber

In some cases the absorber is of the double-pipe type, similar to double-type condensers and brine coolers. The type here illustrated consists of a horizontal cylinder containing coils of pipe through which the cooling water which has previously done duty in the ammonia condenser passes. The cold, weak liquor is admitted at the top of the cylinder, passes down among the cooling coils, and there mingles with and absorbs the cold ammonia gas from the brine cooler.

As the cooling water has already been heated through several degrees in passing through the pipes of the ammonia condenser, it is expedient that a counter-current cooling effect be carried out, in which the warmer, outgoing cooling water cools the weaker aqua ammonia, and the incoming cooler water is employed to reduce the temperature of the strong aqua ammonia on its way to the generator through the ammonia pump and exchanger.

Exchanger

The exchanger is a horizontal steel shell capable of carrying the full generator pressure through which the comparatively cool strong liquor is pumped on its way from the bottom of the absorber to the bottom of the generator. This rich liquor is shown entering the shell of the exchanger at the right-hand side, after which it traverses a spiral pipe coil and finally passes out into the top of the analyzer, shown just above the generator. The exchanger is provided with nests of pipe coils connected in parallel, through which passes the hot weak aqua ammonia, supplied to the manifold at the top of the exchanger by a pipe running from the bottom of the generator. In the countercurrent flow, the hot weak liquor which must eventually be cooled in descending through the pipes gives up a part of its heat to the cooler rich liquor which must eventually be heated when ascending around the coils. By this heat exchange, cooling water is economized in the absorber and steam in the generator.

Analyzer

Leaving the top of the exchanger shell, the cool rich liquor passes to the top of the analyzer. Here it is allowed to trickle down over a set of metal trays, where, coming in contact with the ammonia vapors rising from the generator, the countercurrent heat-exchanging effect is still further continued. The hot ammonia vapor, entraining more or less aqueous vapor as it rises from the surface of the liquid in the generator, in passing up through the analyzer on its way to the condenser encounters the descending rain of cool rich liquor on its way to the generator. The former, which must eventually be liquefied in the condenser, is cooled, and the latter, which must eventually be boiled in the generator, is heated. This advantageous heat interchange also has the effect of actually condensing and returning to the generator a large percentage of the entrained aqueous vapors passing off with the ammonia, and also of liberating some ammonia gas through the heating of the gas-saturated rich liquor. Where ample analyzer capacities are employed, the incoming rich liquor should be within a very few degrees of the evaporating temperature by the time it finally reaches the surface of the boiling liquid.

Generator

The generator, or "still," as it is frequently called, is the ammonia boiler for the evaporation of the weak liquor enriched and changed into strong liquor by the absorption of ammonia gas direct from the expansion coils in the absorber. It consists of a substantial steel shell provided with a heavy cast-iron head through which pass the ends of the steam coils supplying the heat required for driving off the ammonia. The strong aqua ammonia enters at the top, where it remains by virtue of its specific gravity being lower than that of the weaker liquor at the bottom of the generator. The water of condensation from the steam used in the generator is usually returned to the boilers. This can be effected either by an automatically controlled pump, or by a suitable high-pressure trap.

Rectifier

After leaving the analyzer, the hot ammonia gas passes to the rectifier, a water-cooled coil of pipes of sufficient area to insure the condensation of any aqueous vapors that may have escaped condensation in the analyzer. The liquid condensed in the rectifier is rich, saturated liquor which is returned to the generator by way of the analyzer.

Condenser

From the rectifier the ammonia gas, which should now be practically free from aqueous vapors, is passed to the condenser. Condensers for absorption machines, like those for compression machines, may be of either the atmospheric double-pipe or shell type as preferred. The anhydrous ammonia liquefied in the condenser passes to an anhydrous receiver, similar to those used in the compression system, from which it is drawn as required for expansion in the brine cooler, its flow through the feed line being regulated by the usual expansion valve. After evaporation in the brine cooler the ammonia gas again passes to the absorber, after which the working cycle is repeated.

Cycle Traversed by Ammonia

This can be readily traced by following the course of the heavy arrows in Fig. 23. The circuits of both the gaseous and the aqueous components of the aqua-ammonia refrigerant, as well as that of the cooling water, can be more readily followed out by means of the diagram, Fig. 23, in which all mechanical details have been omitted. The several members of the refrigerating system illustrated in Fig. 23 are represented by shaded areas occupying approximately the same relative positions on the diagram. The path of the ammonia is represented by a heavy solid line; that of the water component of the aqua-ammonia refrigerant by a narrow solid line; and that of the cooling water by a broken line. The direction of travel in each case is indicated by arrows.

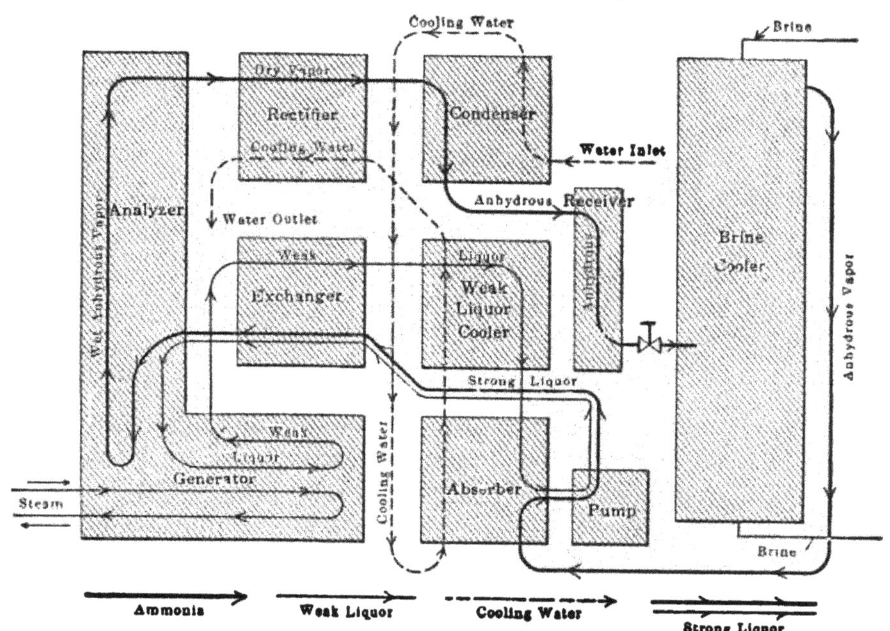

Fig 24.—Diagrammatic Representation of an Absorption Refrigerating System

From this diagram, as well as from Fig. 23, it will be seen that, as "anhydrous ammonia," the refrigerant starting from the "anhydrous receiver" passes to the "brine cooler," where in changing to the gaseous state it performs its sole function of absorbing heat from the brine. As saturated low-temperature ammonia vapor, the refrigerant starting from the brine cooler passes to the absorber, where it enters into solution or is absorbed by the weak liquor from the generator, forming strong liquor. As hot strong liquor the refrigerant starting from the absorber passes through the exchanger, where it gives up some of its heat to the weak liquor on its way to the absorber, then on by way of the analyzer into the generator, where the ammonia gas is driven out of the strong liquor solution, under high pressure, by the application of heat, and passes through the analyzer and rectifier into the condenser, leaving the impoverished aqua ammonia or weak liquor behind in the generator.

In the condenser, the heat originally absorbed by the anhydrous ammonia in changing from the liquid to the gaseous state in the brine cooler, as well as that added to increase its temperature and drive it out of solution in the generator, is given up to the cooling water, circulated through the condenser, causing the ammonia to return to the liquid state, after which it flows to the anhydrous receiver, and the cycle is again traversed.

The aqueous component of the aqua-ammonia refrigerant, starting from the bottom of the absorber in company with the ammonia in the form of strong liquor, passes through the exchanger and analyzer into the generator. Here it is separated from the greater part of the ammonia and returns through the exchanger and weak-liquor cooler to the absorber. Here it again joins the anhydrous ammonia, forming strong liquor, and retraces the path just described.

Path of Cooling Water

The cooling water is admitted first to the ammonia condenser, where it performs its most important function of removing heat from and liquefying the ammonia gas. After leaving the ammonia condenser it is still cool enough to be capable of absorbing a considerable amount of heat from the strong liquor in the absorber, more from the weak liquor fresh from the generator in the weak-liquor cooler, and still more from the hot ammonia gas fresh from the generator in the rectifier, after which it usually passes to waste.

Still another line might have been drawn on the diagram in Fig. 24, indicating the path traversed by the heat from the point of its absorption from the brine in the brine cooler to that of its expulsion with the cooling water from the condenser. Such a line, however, would coincide with that representing the ammonia from the point where the heat and the vapors of the refrigerant leave the brine cooler, continuing to the condenser, where it would cross over and join that representing the cooling water. It would then follow this line through its circuitous passage to the point where, together with the water, the heat flows away to the sewer.

Counter Current Effect

It should be noted that throughout the entire system a counter-current effect is carried out between the cooling and the cooled substances. In the absorber the direction of travel of the cooling water is upward through the cooling coils, while that of the cooled strong liquor is downward around the outside of the cooling coils. In the exchanger, the hot weak liquor from the generator passes in one direction through a nest of pipe coils, while the somewhat cooler strong liquor from the absorber flows in the opposite direction around the outside of the cooling coils. In the double-pipe weak-liquor cooler, the cooling water, after passing through the cooling coils of the absorber, passes in one direction through the inner pipes, while the hot weak liquor passes in the opposite direction between the inner and the outer pipes. In the analyzer the hot ammonia gas passes upward through a rain of cooler strong liquor. Likewise, in the double-pipe condenser a similar counter-current effect is produced.

By these counter-current cooling effects, in which the coldest cooled substance gives up its heat to the coldest cooling substance, and the hottest cooled substance to the warmest cooling substance, the outgoing substance is cooled more nearly to the temperature of the incoming cooling substance than would otherwise be possible, thus effecting economy not only in the amount of the cooling substance required but also in the operation of the system through the reduction in the amount of the refrigerating medium required for a given amount of cooling effect.

CHAPTER V

SIMPLE COMPARISONS

I. Elementary Compression and Absorption Refrigerating Systems

The fundamental natural laws on which artificial refrigeration depends can be readily understood by comparing their action with better known processes. The flow of heat from a higher to a lower thermal level, the operation upon which mechanical refrigeration depends, is clearly illustrated by the flow of water from a higher to a lower level. Heat can no more be made to flow up hill than can water. By the use of appropriate mechanical devices, both heat and water can be raised from lower to higher planes. When water is raised to a given hight, its ability to do work, or its energy, is proportionately increased. When heat is raised to a higher thermal level its ability to do work, or its energy, is also proportionately increased.

Pump for Raising Water

A simple machine for the performance of work on a gas usually takes the form of a pump similar to that used for pumping liquids. Such a machine is represented diagrammatically in Fig. 25. This machine is so constructed that successive strokes of the piston will effect the raising of the liquid from tank 1 to tank 2. To prepare for the comparison which is to be drawn later, let the cylinder be supposed to be filled with sponges, which will not materially affect the operation of the machine. In this case foot-pounds of work are delivered through the piston for the purpose of raising the water from a lower to a higher level and thereby increasing its potential energy. To speak more practically, it transfers the liquid from a place where it is not wanted to some other place where it is.

Pump for Raising Temperature

Only slight modifications are necessary in this mechanism to make it an appropriate machine for illustrating the principle of the compression system as used in the most modern processes of mechanical refrigeration. A refrigerating machine is then a device for removing heat from a place where it is not wanted to some other place where it can be conveniently disposed of. Fig. 26 shows a machine similar in construction to that shown in Fig. 25, but instead of having an inlet and an outlet for the passage of a liquid, its cylinder walls are constructed of a very thin sheet of metal which readily allows the passage of heat whenever a difference in temperature exists between the substance inside and that outside of its walls.

This cylinder employs as a working medium ammonia gas instead of sponges. In this analogy the action of ammonia, or in

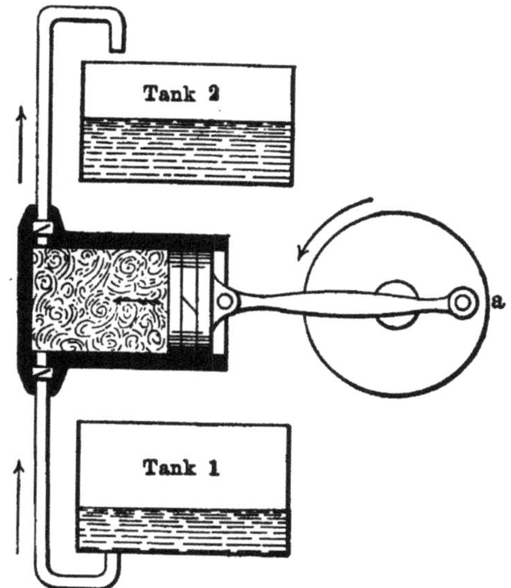

Fig. 25.—Conventionalized Diagram of Pump for Raising Water

fact any refrigerating medium, in absorbing heat, is compared to that of the sponges in absorbing water. After the first charge of heat has been squeezed out of the ammonia and that of water out of the sponges, both the ammonia and the sponges still possess their original capacity for absorbing more. Both the heat and the water are drawn in or absorbed only because of the outside influence exerted by the ammonia and the sponges.

Instead of two tanks at different levels representing difference in "head" or potential energy, in Fig. 26 there are two tanks, *a* and *b*, on the same level, but containing liquids of different temperatures and representing difference in thermal energy. In studying the operation of the elementary refrigerating engine represented in Fig. 26, it must be assumed that the cylinder is to be filled with ammonia gas at a temperature t_1, and a pressure p_1, the piston occupying that position in its path farthest from *X*. If the crankshaft be turned through one-half of a revolution the piston will arrive at *X*. The volume of the gas having become greatly diminished, its temperature and its pressure will have been proportionately increased to t_2 p_2*.

At the end of stroke the piston is allowed to stop and by means of the three-way cock *C*, a spray of water is allowed to

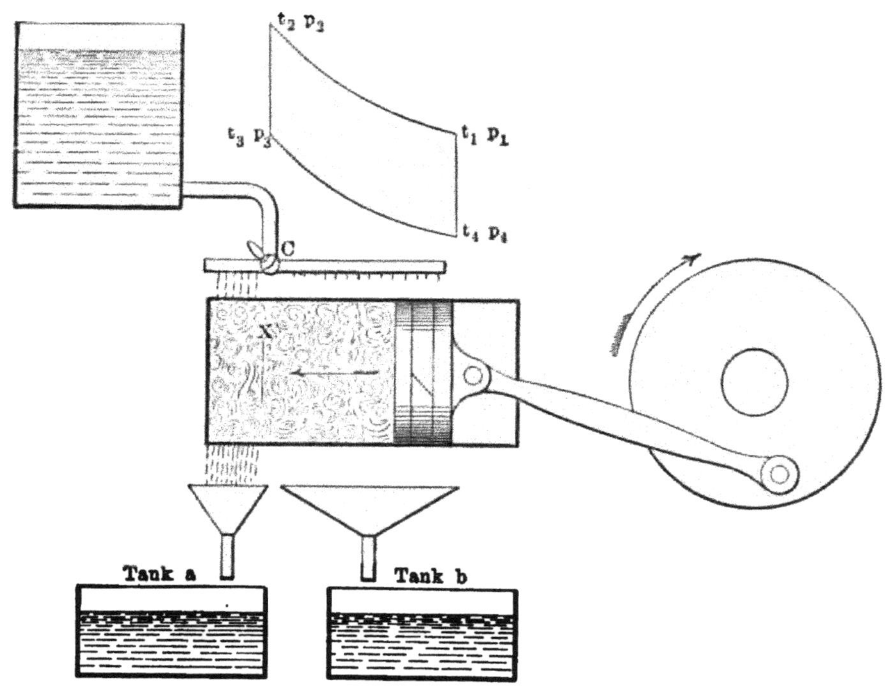

Fig. 26.—Conventionalized Diagram of Compressor for Raising Temperature

flow over the end of the cylinder in which the hot gas is confined, absorbing a portion of the heat of compression, and consequently reducing its temperature and pressure to t_3 p_3. The machine is now made to complete its revolution and the gas, by expanding with the increasing cylinder volume, cools itself to the conditions of t_4 p_4. At the other end of the stroke the piston is again allowed to stop, and by reversing the three-way cock a spray of water is allowed to flow over that portion of the cylinder through which the

* The ratio of pressure to volume in the case of adiabatic compression for ammonia is expressed by the equation $\frac{P}{P_1} = \frac{V_1^{1.3}}{V^{1.3}}$, which means that the pressure will vary inversely in the 1.3 power of the volume.

piston has just passed. The gas within having now become colder than the water, because of having expanded after reaching the temperature of the water, again absorbs heat from it as it passes in a thin sheet over the comparatively large surface of the cylinder, and its temperature and pressure are raised to their original conditions of $t_1\, p_1$, after which the cycle is again traversed.

Working Mediums

In this example the ammonia gas within the cylinder may be compared to the sponges in the preceding case. During the expansion the gas absorbs heat and the sponges water. During compression both the heat and the water are expelled and directed into "places where they can be used, or conveniently disposed of." In the operation of the mechanism represented in Fig. 26, heat is taken away from a part of the water passed over the cylinder and given to the other part. The former, or "refrigerated" water, being conducted into tank *b*, and the latter, which contains a double share of heat, into tank *a*.

In order to more closely approach the mechanism employed in practical refrigeration systems, it will be possible, without in the least altering the principles involved, to slightly modify the system represented in Fig. 26. In practice the operation of the compressor must be continuous and the cylinder proper would have far too little surface to allow of its being used either as a gas cooler or a gas heater, as was done in the foregoing example.

Compressor and Expansion Engine

Let it be assumed, as shown in Fig. 27, that a portion of the cylinder at the top and at the bottom be drawn out, forming a cooling or condensing coil *B* and a heating or expansion coil *D* and at some convenient point, *E*, the passage between the two be contracted to a sufficiently small cross section to maintain a higher pressure in the condenser coil than in the expansion coil, the compressor being in operation. If the two coils be sprinkled with water, or, to conform more closely to practice, if the lower coil be entirely submerged, we will have practically the same conditions as those in the foregoing example. Starting with the piston in the position shown in Fig. 26, the cylinder full of gas is under conditions of temperature and pressure $t_1\, p_1$. By compression these conditions are changed to $t_2\, p_2$, which the cooling of the spray of water over the condenser coil tends to reduce to t_3p_3. In the compound machine, Fig. 28, ammonia gas, or whatever the working medium

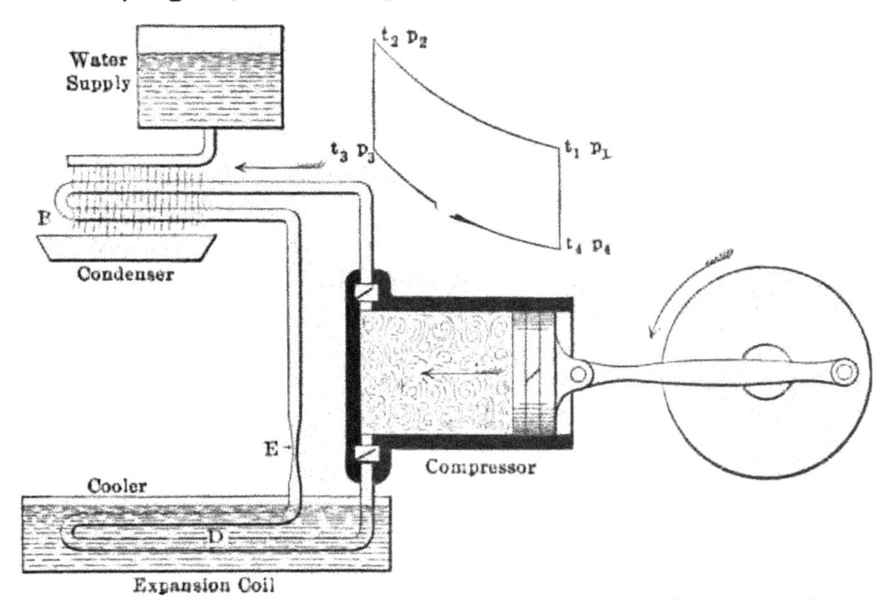

Fig. 27.—Conventionalized Diagram of Compression Refrigerating System

may be, is compressed in the compression cylinder *A*, after which it passes to the condenser *B*, where it is relieved of a large part of its

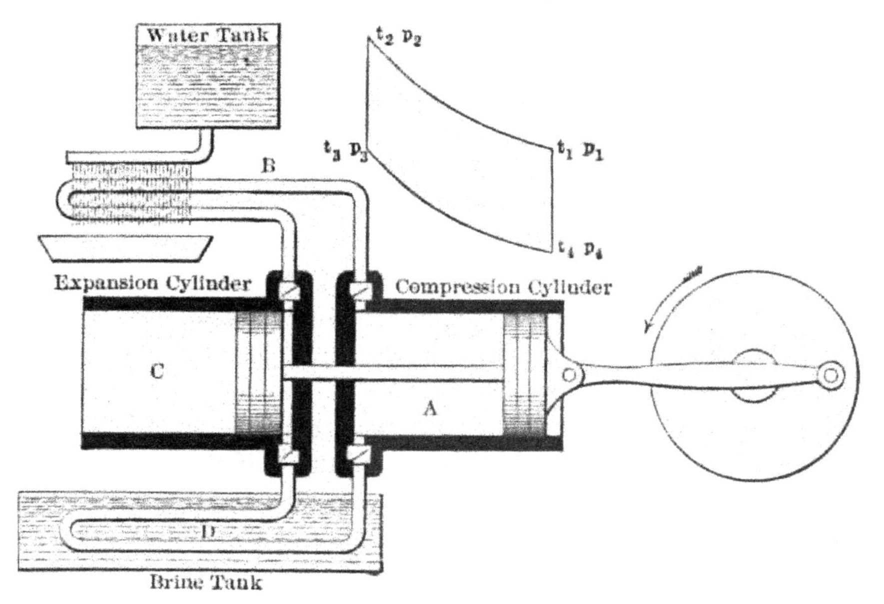

Fig. 28.—Conventionalized Diagram Showing Compression Cylinder for Raising and Expansion Cylinder for Lowering Temperature

heat. From the condenser the gas passes to the expansion cylinder *C*, where it loses more heat because of the work it does in expanding behind the moving piston. This work done by the gas expanding behind the moving piston in *C*, (which is an exact equivalent of the amount of thermal energy lost,) is recovered, since it assists in compressing the gas in cylinder *A*.

REFRIGERATION BY CHANGE OF TEMPERATION OF GASES

The production of refrigeration by the method just described depends on the change in temperature of the working medium, and not on its liquefaction. Machines of this type are, accordingly, not limited in choice of working medium to those which are easily liquefiable, and air rather than ammonia is most commonly employed. Such machines, known as *cold air machines*, have been used extensively in the past on shipboard, their favor there being due to the dangers, real and imaginary, incident to the use of ammonia and other liquefiable refrigerants.

Refrigeration is produced by cold air machines by the process just outlined. The air is first compressed, raising its temperature, and allowed to do work in expanding behind the piston of an air engine, just as steam is allowed to do work by expanding behind the piston of a steam engine. The air, as well as the steam, is cooled, because of the heat converted into mechanical work. The exhaust air escaping from the cylinder of the air engine into the atmosphere of the cold-storage room is often as low as from −50° to −85° Fahrenheit while exhaust steam, though similarly cooled by the performance of work in the steam cylinder, usually escapes to the atmosphere at a temperature of 212° Fahrenheit and upward.

In practice the expanded air from a cold-air refrigerating machine is returned to the compressor and used over and over again, as this reduces the losses in efficiency due to the presence of atmospheric moisture. The production of cold by the expansion of compressed and cooled air is often apparent in the operation of pneumatic machinery in cold weather, in which case the temperature of the exhaust is often low enough to freeze the aqueous vapor, making it necessary to take precautionary measures to prevent the closing up of exhaust outlets by the accumulation of ice. In practical refrigerating systems, in which the working medium is carried through the liquid state in order to make use of the latent heat of liquefaction, the expansion cylinder is omitted as in Fig. 27.

In the example cited in connection with Fig. 26 the available refrigeration is obtained by simply raising and lowering the temperature of the ammonia gas, of which, since one degree of sensible heat represents only one-half a British thermal unit, 288 pounds would have to be lowered one degree in order to produce an effect equivalent to the melting of one pound of ice. This would necessitate handling tremendous volumes of gas to produce comparatively small results of refrigeration. The specific heat of air being only about one-half that of ammonia vapor, approximately twice as many pounds of the former as of the latter refrigerant would have to be employed to produce a given cooling effect. At atmospheric pressure the latent heat of vaporization for ammonia is 573 and the specific heat of the liquid is unity, while that of the gas at constant pressure is only about .5080. Comparing these last values, we naturally look for means of producing refrigeration through the aid of the latent heat of vaporization because of the great heat absorbing capacity of this process. The gas, instead of expanding behind a piston, is allowed to escape through a restricted opening, *E*, Fig. 27, into the expansion coil below, in which a considerably lower back pressure is maintained through the efforts of the compressor which is constantly drawing gas from the lower coil and discharging it into the upper coil. In passing the point *E*, the conditions of temperature and pressure of gas drop to t_4 p_4, which the heating effect of the liquid surrounding the expansion coil finally restores to t_1 p_1, at which point it meets with a repetition of the cycle. The utilization of the latent heat of liquefaction requires no mechanical alterations in the system, and this elementary mechanism illustrated in Fig. 27, when put in operation with a proper charge of anhydrous ammonia and cooling water for the condensers, is capable of producing commercial results of artificial refrigeration.

An elementary absorption machine, corresponding to the compression machine illustrated in Fig. 27, is shown in Fig. 29. As is pointed out in another chapter, the absorber of the absorption machine performs the function of the suction stroke of the compressor in the compression system, and the generator that of the compression stroke. The remaining portion of the cycle, including expansion at E, and evaporation in the expansion coil also remains as in the compression system.

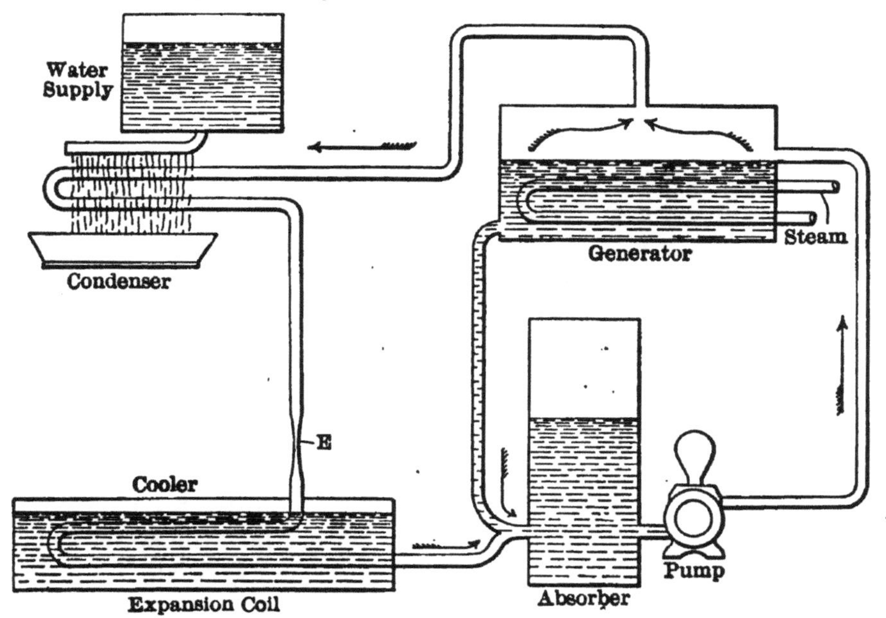

Fig. 29.—Conventionalized Diagram of Absorption Refrigerating System

II. SIMPLE COMPARISONS

Water and Ammonia Systems

It has already been shown that the amount of heat required to raise the temperature of either a solid, a liquid or a gas through one degree is very small compared to that required to effect a change either from the solid to the liquid or from the liquid to the gaseous state.

In the case of anhydrous ammonia, for example, the specific

TABLE I.

COMPARATIVE PROPERTIES OF WATER AND ANHYDROUS AMMONIA

H_2O		NH_3
	Steam and Ammonia Gas	
4500.	Temperature of decomposition	900.
698.	Critical Temperature	266.
.480	Specific heat. Constant pressure	.5080
.346	Specific heat. Constant temperature	.3911
966.1	Latent heat of evaporation. Atmospheric pressure	573.
.0376	Weight per cubic foot at atmospheric pressure	.055
	Water and Ammonia Liquid	
212.	Temperature of evaporation. Atmospheric pressure	—28.5
1.	Specific heat of the liquid	1 to 1.23
144.	Latent heat of fusion	(............)
62.42	Weight per cubic foot at 32° F	42.02
	Ice and Ammonia Solid	
32.	Temperature of fusion. Atmospheric pressure	—115.
.5	Specific heat of the solid	(............)
57.50	Weight per cubic foot at 32°	(............)

heat of the liquid is about unity, that of the gas about .51 under constant pressure and about .39 at constant volume, while the latent heat of evaporation under atmospheric pressure is about 573 B.t.u. For comparison of these characteristics with similar ones for water see Table I.

It has also been shown that refrigeration or cooling is always effected by drawing heat out of one substance into some other substance at a lower temperature. In this case the former substance is refrigerated and the latter heated. In fact *refrigeration and heating are two different views of the same operation.*

Refrigeration and heating may take place at any temperature. Metals, for example, are melted in a furnace. To effect the change from the solid to the liquid state considerable amounts or heat have to be supplied to satisfy the latent heats of fusion, and we may state just as accurately that the melting metals refrigerate the furnaces as we can that melting ice refrigerates a refrigerator.

If the average temperature of our terrestrial atmosphere were a few hundred degrees higher on the thermometric scale, we might actually employ the latent heat of fusion of lead, tin or other metals as we now employ ice, i.e., as a convenient vehicle for absorbing and carrying away comparatively large quantities of heat. With the low temperatures that we can now produce we might at the present time actually employ congealed mercury in our refrigerators in the place of ice were there any advantage to be gained in producing cold by the fusion of a metal. Any substance might be used for absorbing heat for the purpose of cooling other substances. The limits of temperature between which practical use of refrigeration can be made are so narrow, however, that the latent heat of fusion of only a very few substances can be employed and, as a matter of fact, only that of one substance, ice, is employed. Similarly the latent heat of evaporation of only a comparatively few substances falls within that range, and since, unfortunately, none of these can be employed because of other reasons, only the specific heats of available natural cooling media, which as has already been shown afford only limited heat absorbing capacities, can be employed.

Natural Cooling Mediums

Even the so-called refrigerating mediums such as ammonia, carbon dioxide, etc., which substances are liquefiable at such temperatures as to make their respective latent heats of evaporation readily available for the production of artificial cold, must in turn be cooled by some natural cooling medium after they have absorbed their fill of heat in the cold-storage compartment. As will now be shown the most commonly employed natural cooling medium is water. As there are no mediums having their boiling points so located as to make their latent heats of evaporation available for cooling the primary refrigerants such as ammonia, carbon dioxide, etc., we must of necessity employ the more limited heat absorbing capacities available in the specific heat of some convenient natural cooling medium. These heat absorbing capacities available in changes of temperature without change of state are so very limited that only the most inexpen ive substances such as air and water can be considered for cooling on the large scale necessary in the case of steam and ammonia condensers. As a matter of fact, these two agents are practically the only mediums we have that can be employed as primary cooling agents

Air is everywhere available, but often at a prohibitively high temperature. It has the further disadvantage that its specific heat is very small, being only .2377. Since its weight is only .0706 per cubic foot the prohibitive volume of 59.5 cubic feet would have to be employed to absorb a single B.t.u., this assuming a rise of one degree in the temperature of the air, or 16,136,000 cubic feet to carry off 288,000 B.t.u., or an amount equivalent to the negative heat of one ton of refrigeration. Water is more expensive than air, but its specific heat being 1.0 and its weight being 62.5 pounds per cubic foot, only 0.016 cubic feet is required to absorb one B.t.u. Assuming a rise in temperature of one degree, 4,608 cubic feet will carry away 288,000 B.t.u., or an amount equivalent to the negative heat of a ton of refrigeration. Water, after all, is the natural refrigerating medium, the comparatively small specific heat of which we are able to employ only because of the cheapness of the medium.

Where the price of water is abnormally high the still cheaper medium air is employed to cool it, in which case air, with its extremely small specific heat, becomes the cooling medium into which the animal heat, solar heat and heat of fermentation, removed by mechanical refrigeration from cold-storage houses and breweries, finally gravitates. The temperature of the boiling-point and also the latent heat of evaporation of water is unfortunately too high, at the usual working pressures, to enable us to use water as a direct refrigerant; otherwise this medium would undoubtedly be employed in the place of the several mediums now used.* The so-called refrigerating mediums are employed only because it is difficult for water, when used as a refrigerant, to *reach down low enough* to get hold of heat at cold storage temperature levels. These so-called refrigerating mediums are accordingly employed to *splice out* water which is the real arm of every form of *heat dredge.*

Heating and Refrigerating Systems

The striking similarities between a steam and a refrigerating system, together with the fact that almost everyone is more or less familiar with the principles involved in a steam boiler and engine plant, makes such a system most happily convenient for illustrating the principles of operation of a direct expansion refrigerating system.

Except that the cycle of operations is reversed, a steam system consisting of a boiler, an engine, a surface condenser and means of returning the condensation from the condenser to the boiler, is mechanically and thermodynamically an almost exact counterpart of a refrigerating system consisting of expansion coils, a compressor, a surface condenser, and means of returning the condensation from the condenser to the expansion coils. Theoretically ammonia might be employed as a working medium in the steam system for the production of power, and water might be employed as a working medium in a refrigerating system for producing refrigeration. As commercially operated the prohibitive disadvantages of so employing the working mediums are too apparent to warrant comment.

Evaporating and Condensing Members

In these two similar systems a shell boiler has its counterpart in coolers of the shell type and a water tube boiler in direct expansion coils. Each is a receptacle for the evaporation of its respective working medium, the evaporation being accomplished in the for-

* As a matter of fact, as can readily be seen from tables of properties of saturated steam, water can be employed as a working medium for producing refrigeration when not too low temperatures are required. The impracticability of operating an ordinary compression system employing this medium is evident from the fact that to cause water to boil at a temperature even as low as 32° Fahrenheit requires a vacuum of 29.74 inches of mercury.

mer case by the absorption of heat liberated in the combustion of coal in the fire-box, and in the latter by animal or solar heat with which the liquid comes in contact in its journey through the expansion coils immersed in comparatively hot brine or atmosphere of a cold-storage compartment. The system just described is represented diagrammatically in Fig. 31. Corresponding parts of a similarly constructed refrigerating system are shown in Fig. 30.

REFRIGERATING FURNACE GASES AND COLD STORAGE AIR

In the former case heat gravitates from the furnace gases to

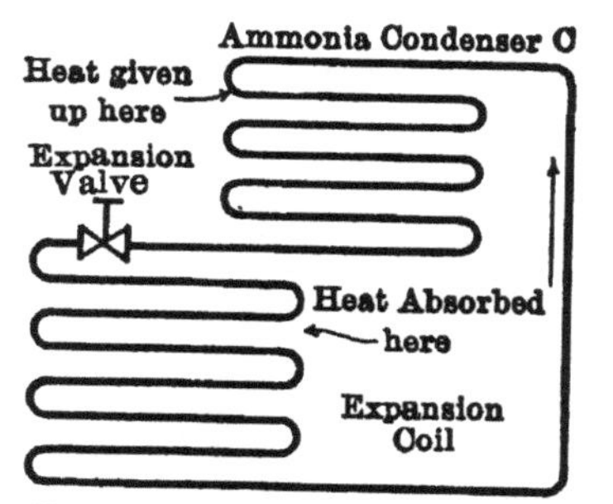

Fig. 30.—Refrigerating System

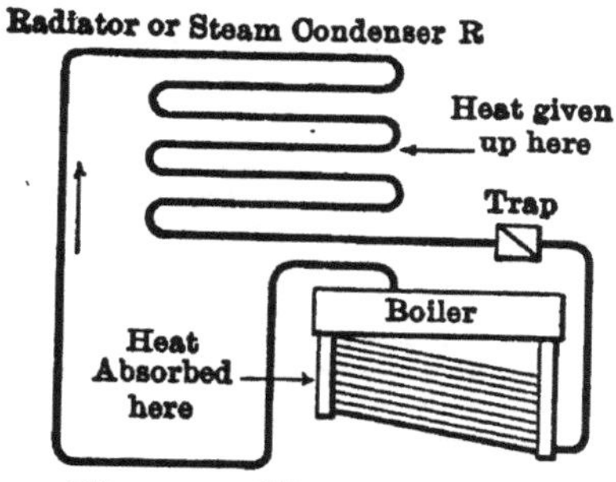

Fig. 31.—Heating System

Diagrams Showing Similarity of Evaporating and Condensing Units of Heating and Refrigerating Systems

the water in the boiler because it is at a lower temperature. The water is heated by the flow and the furnace is refrigerated. The steam generated, we will assume, passes to a steam radiator where heat gravitates from it to the cooler surrounding air. By this flow the air is heated and the radiator is refrigerated. If the surrounding air were hotter than the radiator, it is evident that the heat flow would be in the direction of the outside air to the radiator.

In the latter case heat gravitates from the comparatively hot air of the cold-storage compartment to the ammonia in the expansion coils because the ammonia is at a lower temperature. The ammonia is heated and the air is refrigerated. The ammonia vapor, we will assume, then passes to the ammonia condenser, where, under the same relative conditions as existed in the steam heating system, namely of the surrounding air being colder than the vapors, it would lose its heat and condense. As the temperature of the outside air or even any available cooling water is far above that of ammonia vapors returning from the expansion coils, a condition similar to that of the hot radiator in a still hotter room exists, and the heat flow will be in the direction of the ammonia and no condensation can take place.

Condensation of Steam and Ammonia

So far, the difference in the two systems is that, whereas in the case of the steam system, the vapors are hotter than they need be in order to be condensed by the air in the warm room; in the case of the refrigerating system, the ammonia is not hot enough to be condensed at the temperature of the atmosphere or any available condensing water. The ammonia vapors, as well as the steam, must be condensed, however, before they can be made to absorb more heat in the expansion coils and the boiler, respectively;

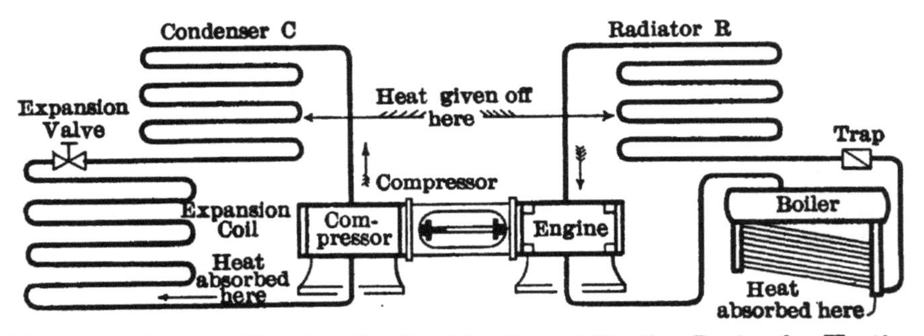

Fig. 32.—Diagram Showing the Combination of Similar Parts of a Heating and a Refrigerating System with a Steam Driven Compressor to Form a Compression Refrigerating System

but as the steam is already hotter than need be it may be robbed of a part of its heat by making it do work by expanding in the cylinder of a steam engine. Since the ammonia vapor is too cold to be condensed by any available cooling medium its temperature must be raised. This is accomplished by doing work on it, a convenient method of which is to pass it through an ammonia compressor. Since there is too much heat on the steam side and not enough on the ammonia side, the two systems may be connected together in such a way that the power developed by the steam is exerted on the ammonia. Mechanically this is effected by means of a direct-acting steam-driven ammonia compressor, illustrated diagrammatically in Fig. 32.

In the absorption system, instead of converting heat into power on the steam side of the system and the power into heat on the ammonia side of the system, a more direct heat interchange between the two sides of the system is effected by bringing the steam and ammonia into close proximity in a *generator*, represented diagrammatically by the two overlapping coils substituted in Fig. 33 for the compressor-driven engine in Fig. 32.

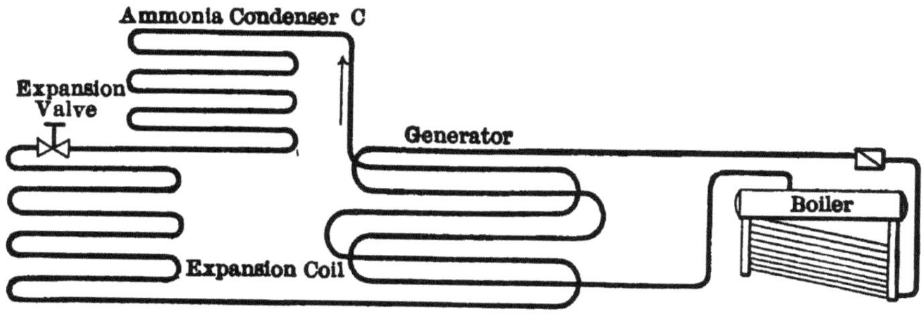

Fig. 33.—Diagram Showing the Combination of Similar Parts of a Heating and a Refrigerating System with a Generator to Form an Absorption Refrigerating System

III. SIMPLE COMPARISONS

The Refrigerating Machine as a "Heat Pump"

In the language of the preceding chapter, the functions of water pumps and heat pumps are to remove water and heat, respectively, from places where they are not wanted to some other place where they can be used or more conveniently disposed of. The rudimentary mechanical systems by which this is effected have been explained. The object of the present chapter is to continue the analogy between the flow of water and that of heat, making a somewhat more practicable concrete application of the principles involved.

Fig. 34, waiving the question of obvious inaccuracy in details, may be taken to represent a sectional view of brewery cellars located on the bank of a river. The floor of the lower cellar is sufficiently high above the river at low water to allow the water from the floors to drain out of a catch-basin sunk in the floor. This condition may be assumed to prevail in the winter, at which time it is also possible, on account of the low temperature of the outside atmosphere, to open the windows and let the heat of fermentation "drain" out. Cold being a condition resulting from the absence of heat, just as dryness is a condition resulting from the absence of moisture, a cold dry cellar can be enjoyed during the days of low temperature and low water without the use of either a refrigerating machine or "heat pump," or a water pump. For emergency, however, the cellars are equipped with both of these pumps, and the walls are waterproofed and insulated as a provision against high water and hot weather. When the river rises sufficiently high the water will flow into the cellars instead of out through the open sewers, and when the outside temperature rises sufficiently high the heat will flow into the cellars instead of out through the open windows. The sewers must be accordingly plugged up and the windows closed. This does not entirely remedy the conditions, however, as the cellar walls are neither perfectly waterproofed nor perfectly insulated and allow both water and heat to percolate in from the outside.

Flow of Water and Heat due to Difference in Level

Water flows from one place to another only when there is a difference of pressure or *static head.* Fifteen feet of water above the level of the cellar floor, since a column of water one foot high and one square inch in section weighs 0.43 pounds, will exert a pressure of 6.45 pounds per square inch in its endeavor to get through the cellar wall.

Heat flows from one substance to another only when there is a difference in temperature or *thermal head.* Seventy-two degrees on the outside of the cellar walls will have an effective thermal head of $72° - 36°$ (the inside temperature) thirty-six degrees, tending to cause a heat flow through the cellar walls.

The rate at which either water or heat will flow from one place to another is directly proportional to the difference in pressure, or temperature, tending to cause the flow and inversely proportional to the resistance opposing that flow. If proper precautions are taken in waterproofing the walls of the cellar, the resistance encountered by the water will be so great that, at the low pressure at which it is trying to gain entrance, there will be no flow. Ordinary precautions against encroachments of heat, however, can at best only reduce, and never entirely prevent, the heat from passing through the walls. While thin sheets of metal, coatings of asphalt and similar materials are absolutely impervious to water there has as yet been no material found which is anywhere near a perfect non-conductor of heat.

Since water always flows from a point of higher to one of lower pressure or level, a convenient means of collecting the seepage in the cellars is by draining it by gravity into a catch-basin still lower than the floors, from which, in the case shown in the diagram, since it is below the river into which it was to be emptied, it must be raised by a pump. The water pumping system (A. Fig. 34) consists of a suction pipe *s*, a pump *P* and a discharge pipe *d*. By the application of power to pump *P* the water in the catch-basin is raised 23 feet to where it is discharged at a height of 7 feet above the river, into which it flows by gravity.

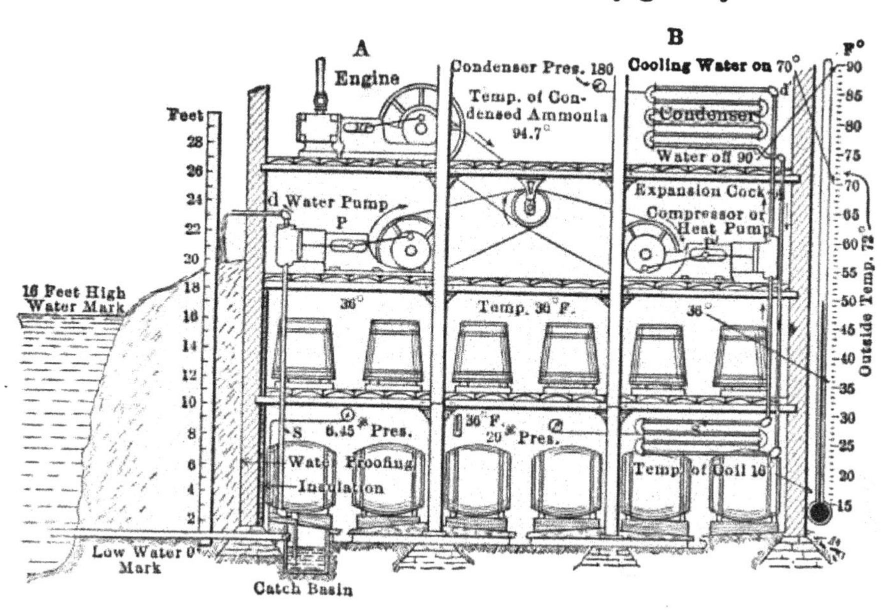

Fig. 34.—Diagram Showing Similarity of Systems A and B for Extracting Water and Heat Respectively

Since heat always tends to flow from a point of higher to one of lower temperature, a convenient means of collecting the heat which filters into the cellar is by *draining it by gravity*, so to speak, into a coil of pipe which is kept still lower in temperature than the cellar. Since this, however, is lower in temperature than any cooling water available for carrying it away (water is generally used for this purpose), its temperature must be raised by means of a heat pump, or compressor. This heat pumping system, *B*, is composed essentially of a low-pressure or suction side *s'*, a compressor *P'* and a high-pressure or discharge side *d'*. By the expenditure of power the gas pump *P'* was employed to raise the temperature of the working medium from that of the coil *s'* to that of the coil *d'*, which is, in this case, about 25° higher than the temperature of the cooling water into which the heat must flow "by gravity."

Working Pressures

The working pressures encountered are fixed by working conditions in the case of both the water pump and the heat pump. The water entering the suction pipe of the water pump is under atmospheric pressure and a sufficient additional pressure must be exerted on it to balance the weight of the column of water 23 feet high or 23×0.43 pounds, about 10 pounds per square inch. If the pump were under the column of water it would have to exert this amount of direct pressure; but since it is located above the column of water it must create a vacuum of 10 pounds, reducing the absolute pressure to 5 pounds, so that the atmospheric pressure of approximately 15 pounds on the water in the catch-basin will exert the required unbalanced upward pressure of 10 pounds.

In the case of the heat pump, the working pressures are determined by the working temperatures, viz., the required temperature in the cold-storage compartment and the temperature of the available cooling water from *A*. The whole problem depends on the temperatures at which the refrigerating liquid will boil under various pressures and the various amounts of heat that must be added to it to evaporate the liquid at the boiling point, or to be abstracted from it to liquefy the vapor at the boiling point.

Evaporating Temperatures

Every liquid has a certain characteristic, fixed boiling temperature or boiling point, corresponding to each pressure. At atmospheric pressure, for example, water, the best known liquid, has a boiling point of 212° Fahrenheit. Alcohol boils under atmospheric pressure at 173° Fahrenheit, ether at 96° Fahrenheit, and anhydrous ammonia, at the same pressure, boils at − 28.5° Fahrenheit. This temperature is quite high as compared with the boiling point of the so-called fixed gases. If the pressure is increased on any liquid the temperature must also be raised in order to make it boil. No two known liquids boil at the same temperature under the same pressure.

Suppose, for example, that there is a pound of liquid ammonia in the coil s', in which there is a gas pressure of 55 pounds by gauge. The temperature of the cellar is 36° Fahrenheit. When the compressor is started it begins to pump gas out of the coil, just as a water pump lifts water out of the catch-basin. Now under 55 pounds pressure the pound of liquid ammonia will not boil because the surrounding temperature is 36° Fahrenheit, which is two degrees less than the boiling point of ammonia at that pressure. As the compressor continues to pump gas out of the coil the pressure will eventually become less. At 52.6 pounds pressure the boiling point of ammonia is 36° Fahrenheit, or the same temperature as the cellar. At this pressure the liquid will be on the point of boiling, but as there is no difference in temperature between the ammonia within the pipes and the atmosphere outside there is no tendency for heat to flow from the atmosphere to the liquid, and without heat it cannot boil. When the pressure is reduced to 50 pounds, however, the boiling point will have been reduced to 34° Fahrenheit and there will be a difference of two degrees, tending to make heat flow from the air of the cellar to the ammonia. This slight difference in temperature would make the process of refrigerating a cellar very slow or else require a prohibitive amount of pipe surface. As a matter of fact, about ten times this difference in temperature is employed in practice. With a cellar temperature of 36° Fahrenheit and a difference in temperature of 20°, the boiling point of the liquid would be 16° Fahrenheit and the pressure required to give this temperature 29 pounds. Under these conditions, from eight to nine feet of 2-inch pipe would evaporate the one pound of ammonia in an hour. In other words, a difference of temperature of 20° Fahrenheit will cause about 545 heat units per hour to flow through the surface afforded by eight to nine lineal feet of 2-inch pipe.

Evaporation of Water and Ammonia

Similar to evaporation of ammonia in pipe evils, is that of the water in a watertube boiler. In both cases the lower the pressure and the hotter the temperature outside, the more liquid will be evaporated per square foot of heating surface in a given time. In the case of the ammonia, the evaporating liquid cools the cellar, and in the case of the water, it refrigerates the fire-box.

While the water pump has but one function to perform, that of raising the water from the catch-basin to the discharge level, the compressor must not only raise the ammonia gas from the expansion coil in the cellar and discharge it into the condenser on the roof *d′*, but also raise the thermal level of the ammonia to a point where its heat can gravitate into the cooling water, which causes the ammonia to return to the liquid form. If the cooling water is supplied to the condenser at 70° Fahrenheit and flows away at 90° Fahrenheit, the condensing pressure, according to the amount of cooling water and cooling surface employed, will be from 180 to 190 pounds, and the boiling point, or temperature at which the gas will liquefy, corresponding to these pressures, will be from 94° to 97° Fahrenheit. The maximum temperature attained by the gas, however, may be very much higher than this.

Condenser water carries away the heat from the cellar, though it is several degrees hotter than the cellar, just as the river carries away the water from the cellar though it is several feet higher than the cellar. The compressor raises the heat through a certain number of degrees, thereby increasing its "thermal head," just as the pump raises the water through a certain number of feet, thereby increasing its "static head."

IV. SIMPLE COMPARISONS

Thermal and Static Head and the Flow of Heat and Liquids

In following out the present comparison it should be remembered that coal is simply the vehicle for bringing us heat radiated by the sun ages ago. By absorption of solar heat a chemical process took place by which carbon-dioxide from the atmosphere was broken up in the plant cells of the vast prehistoric vegetable growths and formed fixed carbon in the plant tissues and free oxygen exhaled into the air. In the process of combustion of coal, free oxygen from the air again combines with the fixed carbon of the plant forming carbon-dioxide, and the long imprisoned solar heat is liberated. Not only is solar heat of former ages made to do useful work through the evaporation of water in boilers, but the solar heat of the present day evaporates the moisture which, precipitated from the rain clouds, collects to form cataracts with power to turn turbines. Not only is prehistoric solar heat used to evaporate water in boilers, but the solar heat of the present could be used in the same way if its rays could be sufficiently condensed and focused, and used directly for the production of steam power, as well as indirectly for production of water power.

It is not altogether unlikely that future generations will see direct solar energy so utilized. The rapid exhaustion of our present fuel deposits, which has even now advanced to a point where the gravity of the situation can no longer be ignored, has already directed some research in that direction. In the present case, however, it is sufficient to assume that the steam under pressure in the boiler has been raised by coal and that the atmospheric vapors forming the clouds have been raised by the sun. In either case energy has been supplied and energy must be withdrawn before the vapors will be condensed to the original form of water. In raising the vapor from the surface of the earth to the clouds a certain amount of energy is expended. In passing the steam from the boiler to the engine there is also a considerable expenditure of energy.

When the two forms of vapor have reached their respective destinations, however, they still possess a considerable energy or capacity for doing work when pressed into service in appropriately designed machines. When the atmospheric vapor has been divested of a sufficient amount of its latent energy it condenses into rain, and were there a suitable machine at hand a large per cent of the foot pounds of work developed by the falling rain could be utilized. Since comparatively small amounts of water are scattered over large areas, man must content himself with utilizing only the foot pounds remaining in the rain after it has been precipitated and collected into larger masses in rivers and lakes, which, though many feet below the rain clouds where it possesses its greatest potential energy, may yet be many feet above the level of the sea from which it arose and may yet, accordingly, have a considerable capacity for performing useful work.

Water Power

In the present example enormous amounts of power are stored in the torrents of water precipitated on the great watersheds that feed the Northern lakes and finally flow down the Niagara River, a part to turn the wheels of industry and a part to dissipate its energy in raising the temperature of the rocky gorge below the falls.

In Fig. 35 is shown a conventionalized machine for utilizing a part of the energy in a small stream of water diverted from the Niagara River above the Falls. While theoretically a modern vertical turbine might have been employed in this analogy, for simplicity of detail and similarity of comparison a bucket conveyor is shown. Water from the duct leading from the river is discharged into the buckets near the top of a sprocket wheel. The weight of the descending water in the upper chain of buckets turns the lower shaft carrying a second chain of buckets so arranged as to elevate the water accumulating in the shaft below the level of the river, and discharge it into a trough near the point of discharge of the upper chain at such a height that it can flow away by gravity into the river.

The power available in any machine is the product of the force acting and the space acted through. In the present example the distance between the point of charging and that of discharging the buckets is about 50 feet, so that every thousand pounds of water discharged will have exerted 50,000 foot pounds, every 33,000 of which per minute is equivalent to a horsepower of work, and every 778 of which is equivalent to one B.t.u. of heat (50,000 pounds per minute = 1.515 H.P. or 64.27 B.t.u.). In this example it is obvious that the higher the point at which the water can be received and the lower the point at which it can be discharged, the more power will be developed.

The amount of power to be expended in raising the water from the shaft depends not only on the number of pounds of water to be raised in a given time, but also on the number of feet through which it is to be raised. If the water is running into the shaft at different levels it is obvious that less power will be required if provision is made for collecting it and conducting it into the conveyor at about the level at which it enters than if it were all allowed to flow to the bottom of the shaft. If the height of the point of discharge be more than just sufficient to allow the water to flow away to the river, foot pounds of work will be unnecessarily expended. Similarly, if the point of discharge of the water from the upper buckets is higher than necessary to enable the water to flow away freely, loss of power will result. Since the water discharged from both sets of buckets must all flow into the river, the points of discharge may be on the same level, as shown, or at different levels, providing both levels are above that of the river.

Pumping Water and Heat

The analogy is apparent. The source of the water available for producing power is 160 feet above the bottom of the shaft, giving it a potential energy which may be said to be equivalent to that of steam having a temperature of 370° Fahrenheit. This steam may be expanded to the lowest pressure, or the heat may

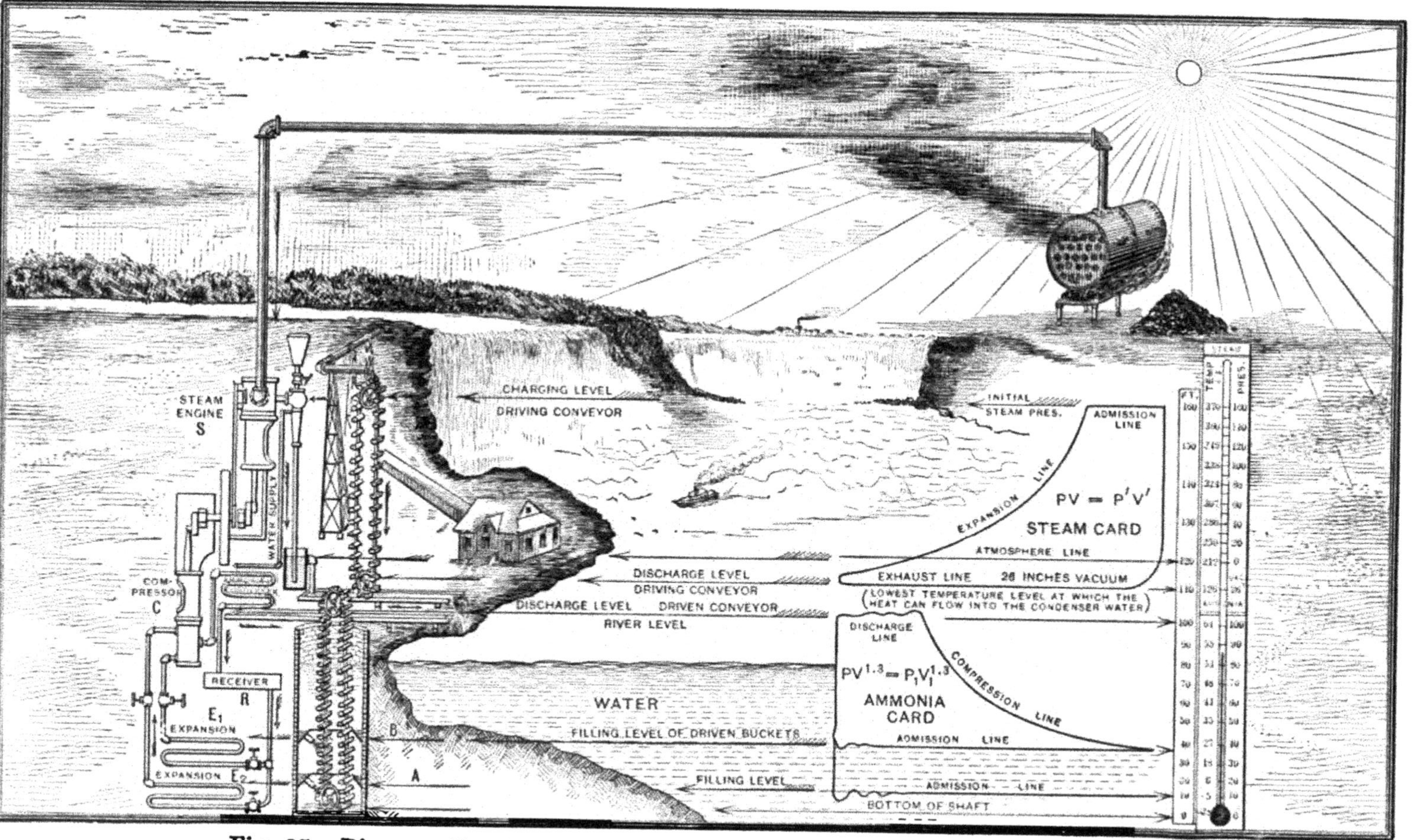

Fig. 35.—Diagram Showing the Similarity of the Gravitation of Water and Heat

be allowed to flow to the lowest temperature at which it can still flow away into the river of condenser water. If this temperature is 126° Fahrenheit the corresponding pressure will be 26 inches vacuum.

This falling of temperature resulting from the conversion of heat into work, in the steam engine shown on the left, is represented by the steam indicator diagram shown on the right, temperatures at any point of which are approximately indicated by the thermometer.

The water to be removed from the pit represents heat to be removed from the lower levels of temperature found in cold-storage compartments. This heat has to be elevated almost to the same level at which heat from the engine is exhausted, because of the fact that the most satisfactory disposition of the heat from both sources is to let it flow into the same river of condenser water. The lower the point of discharge the more power will be available in driving the chain per pound of water, and the less power will be required to raise a given quantity of water by the driven chain. The lower the temperature in the condenser water the more power will be developed in the driving steam engine per pound of steam expended and the less the power required to raise a given quantity of heat in the driven refrigerating machine. In other words, the efficiency of the driving machine depends directly on the difference in head of the water entering and leaving the buckets, just as that of a steam engine depends on the difference in temperature between the steam in the boiler and that in the condenser. Similarly, the efficiency of the driven machine increases directly as the difference in head between the water leaving and entering the buckets decreases, just as that of a compression refrigerating machine increases as the difference in temperature between the gas in the condenser and that in the cooler decreases. Since the pressure of steam at the lowest point to which it is practical to expand it is considerably above that of the refrigerating medium liquefying at the lowest temperature that available cooling water will allow, it is found economical in practice to first permit the heat from the refrigerating medium to flow into the cooling water, after which its thermal level, or temperature, even after being raised by heat from the refrigerating machine condensers, will still be sufficiently low to allow heat from the steam engine condensers to gravitate into it. The diagram illustrates this to the extent of showing that the point of discharge of the water from the driving conveyor is slightly above that of the driven conveyor. This cooling water containing the heat from the steam condenser is shown in the diagram flowing away with the water from the driving conveyor and that from the refrigerating machine condenser with the water from the driven conveyor.

The two sets of expansion coils, E_1 and E_2, located the one above the other, represent the different thermal levels or temperatures at which the heat is absorbed in two cold-storage rooms. The different temperatures at which these two storage rooms are to be maintained is also represented by the height of the spouts which deliver the water seeping through the walls of the shaft into the driven conveyor. The operation of the compressor of the refrigerating machine shown on the left is also represented in the compressor indicator card shown on the right,—heights in feet, temperatures in degrees, and corresponding pressures in pounds, being represented on the three scales also shown at the right.

Leaks

To complete the analogy it is necessary to remember that in the operation of the driving, as well as the driven, chain of buckets there are losses by friction in the bearings as in a steam engine and compressor; losses in capacity due to imperfect filling of the buckets corresponding to imperfect cylinder filling in a compressor; losses due to leaks in the buckets corresponding to leaks by valves and pistons of the steam engine and compressor. In the case of the best steam power plants, all but about 15 per cent of the heat "leaks away" without performing any useful work. In the average steam plant all but about 6 or 8 per cent is lost, so that it is of the utmost importance that this small remaining per cent of heat be utilized to the best possible advantage in the refrigerating machine.

Working Limits

The next most important detail to be considered after that of keeping the compressor in good mechanical repair—that is, to see that the lower buckets do not leak, realizing that, on account of the small per cent of useful work resulting from the expenditure of energy in the prime mover, a given loss in the driven machine is of much greater importance than the same loss in the driving machine—is to see that the condenser pressure or point of discharge of the water is as low as possible and that the expansion coil pressure is as high as possible, or that the buckets pick up the water at as high a level as possible. Assuming that the point of discharge of the water be 100 feet above the bottom of the shaft, 100 foot-pounds of work must be expended in a theoretically perfect machine to elevate a single pound of water. At the efficiency of the average ammonia compressor, from 25 to 35 additional foot-pounds would have to be supplied to make up for leaks and other losses. If the water all enters the shaft at *B*, a point only 65 feet below the point of discharge or 35 feet above the bottom of the shaft and means of directing it into the buckets at this level be devised, only 65 foot-pounds in a theoretically perfect machine, or only from 81 to 88 foot-pounds in a machine of the efficiency of an ammonia compressor, need be expended to do the same amount of work.

Assuming, similarly, that the refrigerating machine, represented by the chain of buckets, discharges the heat at a temperature 100° Fahrenheit above that of the colder refrigerator coil corresponding to the bottom of the shaft, which temperature, for the sake of similarity, may be taken at 0° Fahrenheit, the horse-power per ton of refrigeration would be 1.2194.* Interpolating to find the temperature from which heat equivalent to a ton of refrigeration can be raised by the expenditure of half that amount of power, or 0.6097 horse-power per ton, a cooler temperature of approximately 42½° Fahrenheit is obtained.

If a refrigerating plant is so operated that this heat which enters the refrigerator at such a temperature that it can be absorbed by a refrigerant at 42½° Fahrenheit has to be absorbed at 0° Fahrenheit, in other words, if the plant is operated at 16 pounds back pressure when it could be operated at 61 pounds, one-half the power will be as needlessly expended as would be the case if the water entering the shaft at a height of 50 feet were allowed to flow to the bottom, requiring the buckets to lift it 100 feet instead of 50 feet.

The example just cited need be none the less significant because of the unusual back pressure of 61 pounds gauge. Except for selecting temperatures to agree with the feet head of water

* See table of horse-power per ton of refrigeration—Schmidt, "Compend. of Mechanical Refrigeration," page 449.

already mentioned in the analogy, lower temperatures might just as well have been considered, for example:

The horsepower required per ton of refrigeration when the back pressure is four pounds gauge corresponding to a temperature of 20° Fahrenheit, and the same head pressure of 200 pounds gauge corresponding to a temperature of 100° Fahrenheit, is 1.6090. Again interpolating in the table we find that half this power per ton would be expended when the back pressure is 38 pounds, corresponding to a refrigerator temperature of 24.4° Fahrenheit.

For convenience in comparison the foregoing figures are shown in tabular form in Table *II*.

TABLE II.—TABLE SHOWING CONDITIONS UNDER WHICH A TON OF REFRIGERATION CAN BE PRODUCED BY THE EXPENDITURE OF ONE-HALF THE POWER REQUIRED UNDER OTHER CONDITIONS.

Refrigerator		Condenser		Horse-Power	
Pressure, Lbs. G.	Temperature, F°	Pressure, Lbs. G.	Temperature, F°	Per Ton Refrigeration	Per Cent.
16 61	0° 42½°	200	100°	1.2194 0.6097	100 50
4 38	—20° 24.4°	200	100°	1.6090 0.8045	100 50

While from the foregoing it would seem inexcusable to operate a refrigerating plant at a lower back pressure than actually required to produce the desired temperatures, yet it is probable that not over ten per cent of the plants in commercial operation today are operating under anywhere near the advantageous conditions with regard to back pressure that they should.

The problem becomes less easy of solution as the number of different temperatures increases. Again following out the chain pump analogy, let an extreme case be assumed in which 90 per cent of the water flowed into the shaft at a height of 50 feet from the bottom, and the remaining 10 per cent at the bottom. The theoretical amount of energy required to raise one ton, or 2,000 pounds, of water from the levels at which it runs in would be $2{,}000 \times .90 \times 50 + 2{,}000 \times .10 \times 100 = 110{,}000$ foot pounds. If the water be allowed to flow to the bottom of the shaft, however, it will take $2{,}000 \times 100 = 200{,}000$ foot pounds, or 81.8 per cent more power than by the former method. If 90 per cent of a ton of refrigerating duty be performed at 24.4° Fahrenheit, and 38 pounds back pressure and the remaining 10 per cent of the ton at −20° Fahrenheit and four pounds back pressure, the actual amount of power required will be $.90 \times .8045 + .10 \times 1.609 = .8849$ horsepower; but if the expansion coils for producing both temperatures are connected into a common suction line so that all of the work of refrigeration has to be done at −20° Fahrenheit and four pounds back pressure, the power required will be 1.6087, or, as in the case of the water, 81.8 per cent more than by the former method.

To avoid expending this additional amount of power the same solution presents itself in both cases in question. Two separate machines may be installed, one to perform the more difficult work of raising the lesser amounts of water or heat through the greater distance, and the other to perform the less difficult work of raising the larger amount of water or heat through the lesser distance. Either two compression or two absorption machines may be detailed to the two duties, or, on account of the higher efficiency of the absorption over the compression machine when operating on very low temperatures, the work may be allotted to one compression and one absorption machine; or a unit may be employed of the proper capacity for raising all of the water or heat from the higher level, while a second unit is employed to raise the lesser amounts of water or heat from the lower level to the middle level where the other machine begins to operate.

In refrigerating plants equipped with a single double-acting compressor, or two single-acting compressors, the high and low temperature loads are so proportioned that one end of a double-acting compressor can be made to handle the high temperature load and the other, the low; or in the case of two single-acting compressors, the load may be similarly divided between the two machines, with the additional advantage over the use of one double-acting compressor that, if the loads are not properly balanced, the speeds of the machines may be varied to suit.

Still another way out of the difficulty, and one which avoids many complications arising in the preceding case, is to allow the buckets to run to the lower level and pick up such load as there may be, whether large or small, after which additional load is taken on at the higher level. By this method the load is picked up at whatever level it happens to occupy, and the expenditure of the additional energy required to raise the greater part of the load from the lower level is divided.

Multiple Effect Refrigerating Machine

When applied to the compressor of a refrigerating machine, the method is to admit the low pressure gas returning from the coldest expansion coils directly into the cylinder. When a sufficient part of the stroke has been completed to provide for the low-pressure gas, a secondary suction valve is opened and the higher-pressure gas from the higher temperature expansion coils is introduced. The low-pressure gas is prevented from returning through its suction line by the closing of the low-pressure suction valves, or simply a check valve in the low-pressure line. Every different plant is unquestionably a problem in itself, but whatever the plant, one of the most important questions to be borne in mind every hour of the twenty-four is: *Is the compressor operating at the highest possible back pressure and lowest possible condenser pressure?*

CHAPTER VI

ICE-MAKING SYSTEMS

NEXT in importance to the direct utilization of refrigeration for the cooling of perishable products is that of artificial ice making. While there are a number of systems which may in the future modify present methods, practically all the ice produced to-day is made by either the can or the plate system.

THE CAN SYSTEM

In general, the process of manufacturing can ice consists of immersing cans of water in brine tanks not unlike those employed for cooling brine for brine-circulating systems. First, the specific heat, then the latent heat of the water is given up to the brine, which, in turn, passes it on to the liquid refrigerant, most commonly ammonia.

DISTILLING APPARATUS

Since any impurities in solution or suspension in the water fed to the cans are eventually frozen into the ice, it becomes necessary to use water as nearly pure as possible. The purity of ice, however, is somewhat erroneously judged by its transparency. Impure ice may be almost entirely transparent while, on the other hand, pure ice, except for the presence of air which produces whiteness, may be unsalable because of its opaque appearance. To remove air as well as both organic and inorganic impurities from the water, distilling systems are ususally employed in can ice-making plants. As large quantities of water must be evaporated to make the steam necessary for driving the ammonia compressors and other machinery of an ice-making plant, it follows that the boilers and engine logically constitute a part of the water-distilling system.

HIGH PRESSURE SYSTEM

Fig. 36 illustrates diagrammatically the simple or high-pressure system commonly employed in making can ice. As a steam boiler is virtually a thermal filter which separates out, in the form of incrustation and sludge, most of the impurities brought to it in the feed water, the water supply for an ice plant should be selected with particular care, especially as it often becomes necessary to supply raw "make-up" water to the storage tank when the supply of distilled water runs short.

As shown in the illustration, the exhaust steam from the engine driving the compressor passes first to the grease separator in which it is freed of a large part of the entrained lubricating oil by impinging upon baffle plates. From the grease separator it

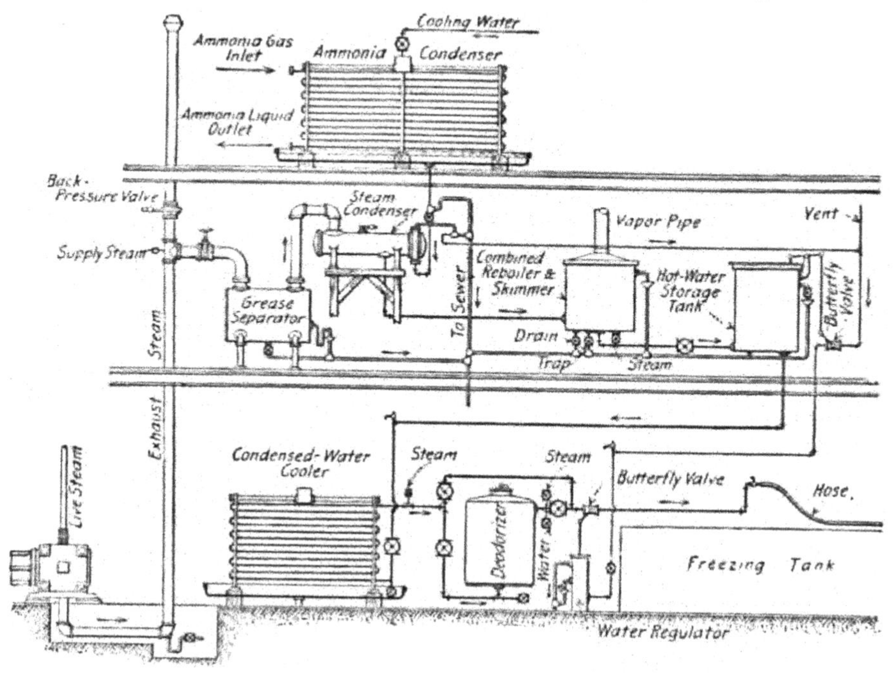

Fig. 36.—Simple High Pressure Distilling System

passes to the steam condenser from whence, after being condensed, it flows to the reboiler, skimmer and hot-water storage tank. From the latter the hot distilled water is allowed to flow as required into the water cooler; entering at the bottom and passing up through a series of pipes it is here cooled by water flowing down over the outside of the pipes. From the water cooler it passes to a charcoal filter or deodorizer and through a hose to the can filler. When frozen the ice is removed from the cans by spraying with hot water, after which it is allowed to gravitate down an inclined chute into the ice-storage room.

In traversing that part of the system between the steam condenser and the ice cans the distilled water, after having been freed from air and other gases in the reboiler, is not again allowed to come in contact with air; the reason for this is twofold: First, any air entering into solution in the distilled water will separate out in the form of minute bubbles during the freezing process and give the ice an opaque appearance; second, distilled water in the presence of air is very corrosive to iron, and should they be allowed to come in contact with any part of the system not thoroughly protected by galvanizing, a sufficient amount of iron would be dissolved to discolor the ice.

Freezing Time Required for Can Ice

With brine at 14 degrees the average time of freezing different-sized blocks of can ice is as shown in Table *III*.

TABLE III. TIME REQUIRED FOR FREEZING CAN ICE

Size of Can, Inches	Weight of Ice, Pounds	Freezing Time, Hours	Size of Can, Inches	Weight of Ice, Pounds	Freezing Time, Hours
6 x 12 x 26	50	15—25	11 x 22 x 32	200	50—72
8 x 16 x 32	100	30—50	11 x 22 x 44	300	50—72
8 x 16 x 42	150	30—50	11 x 22 x 57	400	50—72

While no exact rule can be formulated for expressing the freezing time in terms of difference in temperature between the brine and the freezing water in the can, because of the fact that the heat-transmitting surface of the freezing water is decreasing and the insulating effect of the ice forming is increasing, it, nevertheless, has been claimed by some that the time required for freezing can ice with brine at the usual temperature varies directly as the square of the thickness of the cake of ice. On this basis the relative time of freezing 6-inch and 11-inch blocks would be as 36 is to 121, or allowing 50 hours for the latter, the former should freeze in 14.9 hours.

The Plate Ice System

Where reasonably pure water is available the can system with its distilling apparatus is often replaced by the plate-ice system. The important requisite of any ice-making system from a commercial standpoint is its ability to produce marketable ice, which, unfortunately, often depends more upon the appearance than upon the purity of the product. In the can system practically all solid impurities are left behind in the process of distillation, air and foreign gases being expelled by violent boiling in the reboiler. In the plate system the keeping of the product free from both solid and gaseous impurities is almost wholly dependent upon the agitation of the freezing water. Snow may be pure, but it is white because of the presence of a large number of minute air spaces between the crystals of ice. Gases, in general, are soluble in liquids, the degree of solubility varying widely with the temperature and pressure; the higher the pressure and the lower the temperature, the greater the amount of gas a liquid will absorb. In the case of freezing water, however, the air is driven out of solution and collects in the form of little bubbles on the freezing surface. These bubbles will finally be frozen into the ice if not forcibly dislodged.

Inorganic Impurities

In the manufacture of plate ice the principal inorganic impurities to be guarded against are the salts of iron which give a reddish discoloration, and the carbonates and sulphates of lime and magnesia which produce a slight cloudiness. Unless large quantities of magnesium carbonate or carbonate of iron are present the effects of these impurities, as well as that of air, can be overcome by increased agitation. In the case of carbonates of either magnesia or iron, increased air agitation may tend to increase the discoloration through the hydrating of the former and the oxidizing of the latter. This difficulty may be overcome, however, by the substitution of mechanical for air agitation.

Operation of Plate System

Mechanically, a plate plant is so constructed that the raw undistilled water to be frozen is brought in contact with plates of sheet metal bolted to either brine or direct expansion coils, in which a sufficiently low temperature is maintained to bring about the necessary heat transfer from the water at 32°. These plates, which are not usually less than 14 feet long by 10 feet deep, are submerged in the plate tanks. The refrigerating agent, whether brine or ammonia, is allowed to flow through the coils until ice has accumulated to a thickness of 12 to 14 inches on the plate. The cold brine or ammonia is then shut off and hot brine or ammonia is circulated through the coils until the ice is loosened from the plate and floats free in the water. Chains are then fished around the cake and it is hoisted from the tank by a traveling crane and carried to a tilting table, where it is carefully deposited to avoid breaking. Here it is sawed into cakes of the required size, by two gangs of traveling circular saws, one traveling lengthwise and the other crosswise of the table. Because of being frozen from water at 32° Fahrenheit, with which it is always in contact, the actual temperature of plate ice is not as low as that of can ice, the temperature of which is limited only by the temperature of the brine. On account of this fact, plate ice is less likely to be brittle, has less tendency to freeze together and, therefore, can be stored more readily than can ice.

Center-Freeze System

The factor which limits the application of the plate system is the time required for freezing, the absorption of heat having to take place in this case wholly from one side while that in a can is from five sides. This disadvantage has been overcome to some extent in the "Center-freeze" system, in which the plate of ice is frozen on a comb-formed series of vertical brine pipes attached at the top to suitable feed and return headers. In this system heat is absorbed radially from all directions by each pipe and, being located in the center of the plate of ice, the total thickness of ice frozen through need never be over half that of the usual plate system producing a plate of the same thickness. When the plate of ice is frozen to the desired thickness it is melted loose from the freezing pipes as in the preceding case.

It is claimed by the promoters of this system that the time of freezing is considerably less than half that of the usual plate system operating with brine of the same temperature. As a matter of fact, it would be expected to be less, in the ratio of the square of the thicknesses frozen through, were there the same amount of heat-absorbing surface in each case.

In other words, where 130 hours is required to freeze plate ice by the usual method, it would be expected that, on the basis of equal cooling surface, only about 33 hours would be required. As a matter of fact, with the usual surface and zero brine it is claimed that 11 inches of plate ice can be frozen by this method in 36 hours.

Evaporators and Vacuum Distilling Apparatus

In can ice-making plants of over ten tons daily capacity and employing engines of the Corliss type, there is seldom sufficient sweet or distilled water from the exhaust-steam condensers to supply the freezing tanks. This prohibits maximum steam economy where the usual high-pressure distilling apparatus is employed. For instance, assuming a 100-ton ice plant requiring 2.75 horsepower per ton and operated by a four-valve engine using 30 pounds of steam per horsepower per hour, the steam required for the engine would be about 100 tons per 24 hours. The auxiliaries and reboiler, together with the usual condensation and leakage past the valves of the engine and auxiliaries, would probably amount to 25 tons, making in all about 125 tons of steam. This amount of steam would supply sweet water for the 100-ton can plant and allow for a loss of 20 per cent between the exhaust pipe and ice cans. If, however, the loss were as much as 24 per cent, which might readily happen, the make-up water required would amount to about 8,000 pounds per 24 hours. This quantity of steam at a cost of 20 cents per thousand pounds would be worth $1.60 per day.

If the engine employed were of the Corliss type, simple, non-condensing and having a steam consumption of 28 pounds per horsepower-hour, the steam required to drive the compressor would be about 92.4 tons. Hence, the amount of make-up water, even on the basis of 20 per cent waste, would be 13,200 pounds, which at 20 cents per thousand pounds would cost $2.64 per day.

From the foregoing it is obvious that to employ engines of lower steam consumption results in developing an ice-making capacity in excess of the amount of sweet water available. This excess capacity over that required to freeze the available distilled water may be employed to freeze ice in a plate tank, or the deficit in sweet water necessary to supply the can plant may be made up by means of evaporators.

Combined Can and Plate Ice Plant

A combination can and plate plant, designed to satisfy the first of these conditions, is illustrated diagrammatically in Fig. 37. Leaving the ammonia compressor, the gas is first discharged into the two pressure tanks where any entrained oil is deposited. From there it passes through pipe *B* to the condensers and after liquefying it flows through pipe *D* to the liquid receiver. The line from the bottom of the liquid receiver branches off, line *F* supply-

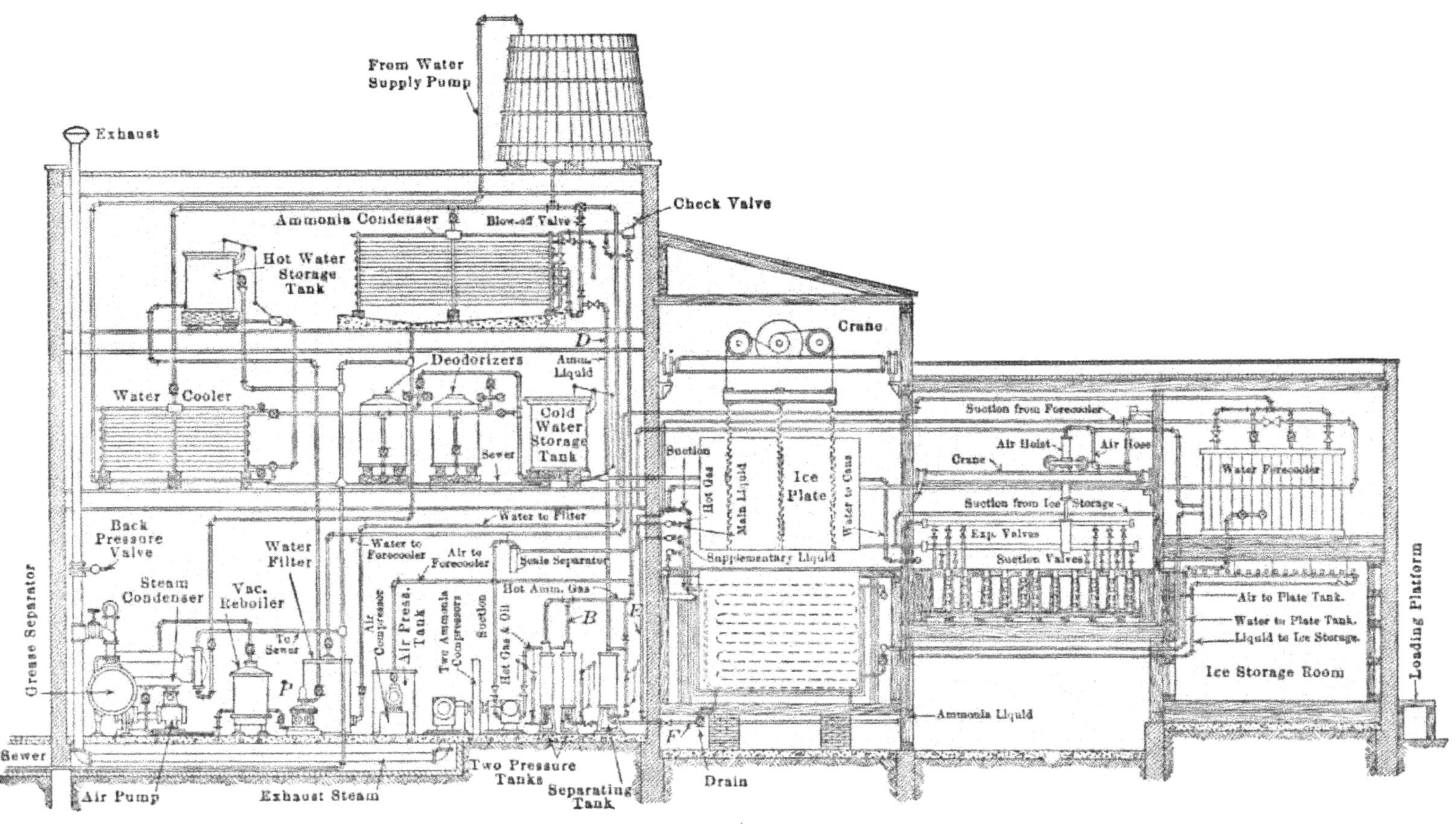

Fig. 37.—General Arrangement of Combined Can and Plate Ice Plant

ing the can plant and ice-storage room and *E* supplying the plate plant. The water forecooler is fed in series with the plate plant, after passing through which, the ammonia gas returns to the compressor.

The circuit traversed by the sweet water is as follows: The exhaust from the engine, encountering the back-pressure valve on the main exhaust pipe from the engine, is diverted through a grease separator into a steam condenser. The condensed water then passes through the vacuum reboiler and enters the suction

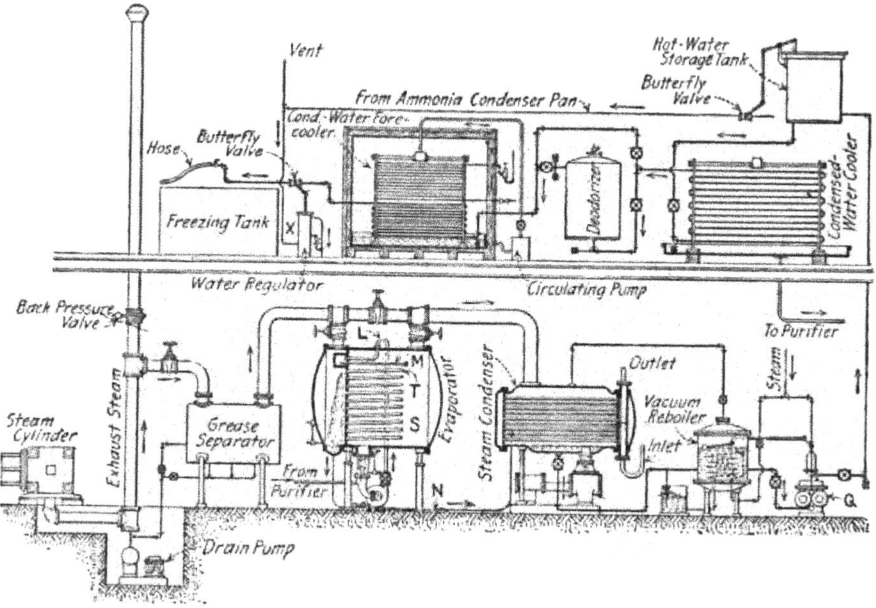

Fig. 38.—Vacuum Distilling System

of pump *P*, which discharges it into the hot-water storage tank; from here it flows through a regulating valve through the water cooler and into the cold-water storage tank, from whence it is drawn to fill the ice cans as required. The water for the plate plant passes first through the water filter in the engine room, through the water foreçooler and into the plate tank. Similarly, the air used for agitation in the plate-ice tank is discharged by the air compressor through an air-storage tank in the engine room, through the water forecooler and into the plate-ice tank.

Vacuum Distilling System

Fig. 38 represents a vacuum-distilling system, having an evaporator which provides the second means of maintaining the full capacity of the ice plant when the available sweet water is insufficient. For simplicity only the distilling part of the ice-making plant is shown in this illustration.

The exhaust steam, as before, passes first through a grease separator, but in this case it also passes into an evaporator, where the steam must stop, the heat being carried over by the vapor to the steam condenser. Assuming that the engine is running under 18 inches of vacuum, the exhaust from the low-pressure cylinder will enter the evaporator at about 168° Fahrenheit. The steam enters the dead-ended copper tubes *T*, which extend upward at a slight angle through the tube sheet into compartment *S*. Here it is condensed by cooling water circulated from the bottom of the evaporator, through the centrifugal pump, distributing pipe *L* and discharge line *M*. On the condenser side of the tube sheet a vacuum of from 24 to 26 inches is maintained by the condenser and this higher vacuum enables the heat liberated by the condensation of every 1.15 pounds of exhaust steam to evaporate about one pound of cooling water. The cooling-water vapors are liquefied in the steam condenser. Here they are joined by the condensed exhaust steam from the evaporator, which is drawn through pipe *N* by the higher vacuum in the condenser, and also by a small amount of vapor drawn through the vent pipe from the top of the vacuum reboiler, condensed water from both the evaporator and steam condenser being drawn into the reboiler by the vacuum maintained in the steam condenser. The water from the steam condensed in the coils of the reboiler drains into a trap provided with a float, which as soon as the water has collected to a certain level, admits it into the suction line leading to the sweet-water pump *Q*. This pump discharges the sweet water into a hot-water storage tank, from whence it flows through the condensed-water cooler, deodorizer and condensed-water forecooler to the ice cans. In the reboiler a float valve controls the operation of the condensed-water pump, allowing it to draw water from the reboiler only when it has accumulated to a predetermined height. In the trap from which the water condensed in the coils of the reboiler is drawn there is a similar float valve, opening only when there is a certain amount of water present. A float valve in the hot-water storage tank controls the position of another valve through which water from the ammonia condenser pan flows into a regulating device *X*, which operates a butterfly valve in the sweet-water supply line leading to the ice cans and prevents the drawing of water from the storage tank below a certain level. These precautionary measures are all taken to prevent the possibility of air entering the pipes of the distilling system.

The deodorizer, into which the sweet water is introduced through a strainer to insure uniform distribution through the filter bed, consists of a vertical cylindrical shell filled with charcoal covered with a second strainer which prevents any of the material from floating and entering the discharge pipe at the top. By means of a simple bypass the deodorizer can be cut out of the system for cleaning, and the sweet water fed direct from the cooler to the cans. In some instances the presence of iron salts in the water makes it advisable to supplement the deodorizer with a sponge filter.

The forecooler shown in the illustration consists of an insulated compartment in which a direct-expansion coil is installed over the distilled-water coil. Water is circulated from the pan beneath these coils and passes over the expansion coil where it is cooled to practically 32°; it then gravitates down over the water coil and absorbs heat from the sweet water. As the circulating liquid is water, it is impossible to freeze up the sweet-water coils, and since this circulating liquid can be chilled to the freezing point the condensed water can also be cooled to within a very few degrees of the freezing point, resulting in a great saving in freezing time, which is equivalent to increasing the capacity of the ice-freezing tanks. The reboiling of the sweet water under a vacuum at a temperature of from 200° to 204° not only reduces the amount of steam required to effect the reboiling, but also the amount of cooling necessary to reduce its temperature to the freezing point.

CHAPTER VII

THE INSTALLATION AND OPERATION OF REFRIGERATING SYSTEMS

Installation

It is scarcely necessary to state that every detail connected with the installation of the piping for ammonia, or other gases employed as a working medium, should be executed with the greatest care. Only such materials as have been found by responsible builders to be well adapted to their respective purposes should be employed. For ammonia, good full weight wrought-iron pipe is to be recommended for the expansion or low-pressure side, and extra heavy pipe and ammonia fittings of approved design for piping the compression side.

In damp places, where the low-pressure gas headers are liable to rust abnormally, these also should be of extra heavy pipe.

Ammonia pipe joints are usually made up with lead or rubber gaskets in male and female flanges, sweated on the pipes. Some builders employ a litharge and glycerin cement in making up screwed joints instead of solder, and there is little difficulty in making such joints tight if scrupulous care is exercised in seeing that only true, sharp, properly formed threads are used; that they are thoroughly cleaned, for which purpose gasolene is to be recommended; that the litharge is freshly and thoroughly mixed into a thin paste; and that the joints are made up tight.

It seems trite to suggest that gasket and flange joints should be drawn up squarely, but many a charge of ammonia has been lost through lack of attention to this detail. Rubber gaskets are particularly likely to blow out of improperly drawn up flanges months later when the rubber has become softened by the oil.

When the erection of the plant is complete, and the piping thoroughly blown out to free it from dirt, scale, metallic chips and other foreign substances, an air pressure of not less than 300 pounds should be pumped on it to test for leaks.

Leaks resulting from split pipes and improperly made up gasket joints may be readily located by the sound. In fact, in a still cooler, sound is the most efficient means of detecting very small leaks, especially of ammonia, when the air has become so laden with the fumes as to make the usual methods of testing difficult. When all of the leaks have apparently been stopped, it is advisable to pump a pressure on the piping and let it stand for ten or twelve hours. The drop in pressure, provided there is no appreciable change in temperature, will indicate the amount of leakage. As a final precaution the air may be allowed to escape and the system again charged with air into which a sufficient amount of ammonia has been fed to make any leak easily detected either by smell or by means of sulphur sticks.* The approximate location having been found by means of the sulphur fumes, the exact position of the leak may be located by oil applied to the leak by means of a long-nosed oil can or soapsuds applied with a brush.

Where there is a likelihood of existence of leaks in pipe or submerged condensers, or other places where escaping ammonia would not readily be detected because of its entering into solution in the cooling water or cooled brine as the case may be, it is advisable to test these liquids periodically with some reliable reagent. Where there are not too many foreign substances present, the litmus and turmeric papers are fairly reliable. A more satisfactory reagent, however, for use under the varied operating conditions of refrigerating and ice-making plants, is Nessler's solution, a few drops of which added to the suspected water or brine will show a yellow discoloration for slight traces of ammonia, increasing with the amount of ammonia present until with large quantities a reddish-brown precipitate is formed.

Repairing Leaks

Many small leaks such as occur in ammonia fittings may be stopped by the judicious use of a set of small calking tools.

Porous spots in iron and steel castings may sometimes be remedied by the judicious use of some rusting solution such as sal ammoniac or hydrochloric acid. Where the leaks are occasioned

* Sulphur sticks used for testing for ammonia are made of pieces of white pine, or other wood which burns with little smoke, split into splinters half the size of a lead pencil and from 6 to 8 inches long. These sticks are then dipped into molten sulphur so that about 4 inches of the ends are thoroughly coated, and after being cooled are ready for use. In testing for leaks, the sticks are ignited and held close to the suspected pipe or fitting. If there is escaping ammonia, it will, on coming in contact with the burning sulphur, produce a very noticeable white cloud.

by blow holes of considerable size occurring where the application of pressure will tend to drive the substance into the porosities of the iron, some of the patented rust-joint preparations may be effective.

Troublesome leaks due to imperfect welds in the seams of pipes may be effectively repaired by first cleaning the pipe with a file and some suitable soldering solution, then applying a closely laid course of bright steel wire. The layer of wire should then be saturated with the soldering solution and the whole surface thoroughly coated with solder, special care being taken to see that it is thoroughly sweated in at the point where the leak occurs. The steel wire supplies the tensile strength, the lack of which in the solder would often allow the ammonia under pressure to lift off the solder coating. A hard solder should be employed and the steel wire should be thoroughly "tinned" to protect it from rust.

For soldering iron and steel pipes two soldering solutions should be employed, the first being simply a cleaning solution of concentrated hydrochloric acid, and the second a saturated solution of zinc chloride, commonly known among tinners as "cut acid." This second solution is prepared by dissolving metallic zinc in concentrated hydrochloric acid. Some builders add to the zinc chloride thus formed an equal amount of ammonium chloride.

Leaks, both in pipes and castings, may be repaired and separate pieces of pipe may be welded together to form continuous pipes, coils and headers, by means of improved processes of electric and oxyacetyline welding. Ammonia receivers, as well as larger shells such as are used for constructing absorbers, condensers and generators of absorption machines, are also made by this process.

Charging a Refrigerating System

After it has been found that the system is perfectly tight, the air and the ammonia should be allowed to escape, after which the whole system should be pumped down to a vacuum.

Even pumping a vacuum does not insure the expulsion of all the air, but it becomes greatly rarified as it expands under the reduced pressure and the remainder may be allowed to stay in the system until displaced by purging at the condensers, or a large percentage of it can be driven out of the system by the judicious manipulation of the ammonia. If, for example, the ammonia be admitted very slowly at one end of a long run of pipe, it will drive the air before it without mixing with it to any great extent, and, if at the other end of the pipe line a valve be opened or a flange union be cracked after sufficient ammonia has been admitted to produce a pressure above that of the atmosphere, the air can be allowed to escape until it contains too large a percentage of ammonia, when the opening is closed.

To initially charge or recharge the system, connect the shipping drums of anhydrous ammonia, one at a time (or more if the plant is of large capacity or the initial charge is being put in and one wishes to save time) to the charging valve usually placed between the master expansion valve on the liquid line, where it leaves the receiver, and the expansion coils or brine cooler. This connection is most easily made by a special fitting built up with two swing joints, one end threaded to fit the valves on the shipping drums and the other provided with a flanged or threaded end to connect to the charging valve. When the connection has been made the air in the pipe may be expelled by slightly opening either the charging or the shipping-drum valve and loosening the flanged swing joint nearest the opposite end.

The connection having been carefully made, the main valve on the receiver is closed and the low-pressure side is "pumped down" by allowing the compressor to continue operation after the liquid has been shut off. By the "pumping down" process the ammonia in the expansion side of the system is compressed and discharged into the compression side of the system, where it is condensed and flows to the liquid receiver which it may fill as well as the lower pipes of the condenser.

When the low-pressure gauge indicates that the pressure in the expansion coils has been reduced to zero, or atmospheric, pressure, the charging valve may be opened wide and then the valve on the shipping drum may be "cracked," allowing a small stream of the liquid to pour into the system. The valve on the drum virtually becomes the expansion valve of the system and its manipulation should be governed by the same rules that govern the other expansion valves when the machine is in normal operation, except that it is better not to carry the back pressure quite as high as usual. This pressure may be anything above atmospheric, but there is an advantage in not reducing the pressure below atmospheric as the vacuum would tend to draw air into the system through the charging connection when the drum is disconnected if the charging valve is not absolutely tight, and a considerable inrush of air is obviously less easily detected than a very slight leak of ammonia outward. While the production of a lower pressure within the refrigerating system, than that of the atmosphere without, undoubtedly hastens the operation of charging, there is always the tendency to draw air or water into the system. A vacuum should, accordingly, never be pumped until it has been demonstrated beyond all reasonable doubt that the system contains no leaks. Small leaks into the system are not readily detected, and it is evident that much more trouble can be made by what water a small leak will let into a system than by the amount of ammonia the same leak will let out.

When an open connection is made between the shipping drum and the system, the liquid is forced out of the drum into the system by the pressure of the gas above the liquid just as water is forced out of the blowoff of a boiler by the steam pressure above the water. The only difference is that it requires a higher temperature than that of the atmosphere in the engine room to raise steam pressure, while any temperature above zero will give a pressure above atmospheric in the case of ammonia.

This is made mechanically possible by the construction of the shipping drum valve, which, after passing through the head, turns down to within about half an inch of the side of the cylinder. When, as is advisable, the opposite end of the drum is elevated a few inches, there remains only a very small volume of the drum below the level of the outlet, and this has its advantage in that heavier impurities tend to remain in the drum instead of passing into the system.

That the liquid ammonia will pass from the shipping drum into the system without the necessity of pumping is evident. If the engine-room temperature be 80° Fahrenheit, for example, the pressure in the drums will be 140 pounds gauge, or there will be 140 pounds difference in pressure between the ammonia and the atmosphere to cause the flow. If the drums are exposed to the sun, the temperature may rise much higher than that of the surrounding air and even dangerous pressures may result. It is accordingly advisable to store ammonia drums in a cool place. The shipping drums are designed to carry any reasonable pressures, but there is a remote possibility that the drum may be filled too full.

In this case, since there is not sufficient vapor space to take care of the expansion of the liquid as the temperature increases, and since liquids are practically incompressible, there is no limit to the amount of pressure that may be produced except that of the ultimate strength of the drum. There is the same danger in tightly closing the valves on all the outlets to the liquid receivers when it is not definitely known that they are not completely filled with liquid. Explosions due to such causes are second only to boiler explosions in their disastrous results.

It is obvious that the vapor generated in the drum will drive the liquid out into the system so long as the temperature of the liquid is such as to produce a gas pressure higher than that in the system. If, for example, the system is operating under 15 pounds gauge back pressure, 16 pounds vapor pressure in the drum would suffice to expel the liquid. The temperature of the liquid corresponding to a pressure of 16 pounds is about 0° Fahrenheit. From this it will be seen that the only disadvantage of charging against back pressure is that the liquid will not flow so rapidly into the system because of the decreased difference in pressure. The slightest reduction in pressure within the drum, due to a removal of part of the liquid, causes the ammonia to boil more vigorously, generating more vapor to fill the increasing space above the liquid. The temperature at which the liquid boils gradually drops, however, until at a pressure of about 47 pounds, which corresponds to a temperature of a little less than 32° Fahrenheit, the pipe leading from the drums will become sufficiently cold to precipitate and congeal atmospheric moisture, and is said to "frost." The melting of this frost when the pressure is 47 pounds or less indicates that there is no more ammonia passing through the pipe. The valves can then be closed, the empty drum removed and a full drum connected.

Amount of Ammonia Charge

It is easier to form an opinion as to the amount of ammonia that the system needs while it is operating than it is to determine when a sufficient amount has been added. Except in the case of initial charges, it is better to add a comparatively small amount of ammonia and then to operate the system for a sufficient length of time to restore normal conditions. The height of the liquid in the gauge glass of the receiver, or the general performance of the plant when no gauge glasses are used, will give the engineer an idea as to whether or not more ammonia is required. It should be remembered that refrigeration is produced by the absorption of the heat required to change the liquid ammonia to a gas, and since it takes only a very small amount of heat to raise the temperature of any gas that passes the expansion valve in company with the liquid, little cooling effect can be expected from the gas.

So far as the production of cold is concerned, there need be only sufficient liquid refrigerant to insure a solid stream at the expansion valves, so that no gas may enter the expansion coils at any time. The passage of gas can be readily recognized by the intermittent hissing sound produced by the passage of quantities of liquid and gas. The condition in which there is just enough liquid to give a solid flow at the expansion valve is the minimum charge that can be economically employed. A lesser quantity must result in loss of both capacity and efficiency.

To provide for unforseen contingencies, such as losses of ammonia through leaks, temporary trapping of liquid in low parts of the system, etc., it is always expedient to have the liquid charge somewhat in excess of this amount, a kind of credit balance in the bank to guard against the embarrassment of overdrawing one's account if collections do not come in from the expansion coils, condensers, etc., as expected.

Increasing the charge of anhydrous ammonia above that actually required to insure an uninterrupted flow at the expansion valve, will work no harm to the system up to the point where the liquid fills the receiver and begins to encroach upon the condensing surface. To be sure, the more liquid lying in the compression side of the system under the usual conditions of operation, the less space there will be for the storing of additional anhydrous ammonia should it become necessary to "pump out" the low-pressure side. The additional ammonia occasions an additional investment, but in the majority of plants it is expedient to carry a small stock of liquid to provide for contingencies; and aside from the rather remote possibility of an accident that would result in the loss of the entire charge, it is better to have the ammonia in use in the system than lying idle in shipping cylinders.

An overcharge of ammonia in a system can usually be detected in two different ways. Since the condensed liquid soon becomes several degrees colder than the uncondensed gas, the parts of the compression side that contain liquid ammonia, whether the liquid receiver connecting piping or pipes of the condenser, can usually be determined by their lower temperature. Assuming that the plant is overcharged to such an extent that the liquid in the compression side begins to fill up the condenser, thus encroaching on the available condenser surface, a material rise in head pressure over that usually observed when running under similar conditions of speed, back pressure and water supply with only sufficient liquid in the system to insure a solid flow through the expansion valve, would be expected. As it is usually impossible to say whether these conditions are exactly constant or not, a slight increase in head pressure observed on increasing the charge should not be accepted as proof positive that the system has been overcharged, even if the increased pressure seems to occur under constant conditions of operation.

There can be no fixed rule by which to determine the amount of ammonia required for a direct-expansion refrigerating system. For systems not including sharp freezers, the only accurate way is to calculate the amount of the charge, taking as a starting point the amount of pipe to be filled with the refrigerant in both the high- and the low-pressure sides of the system. The following tables, showing cubical contents of pipes and weights of gas at different pressures, will be found convenient when calculating the amount of ammonia required to charge the system.

TABLE IV.—RELATION OF CUBICAL CONTENTS TO RUNNING FEET IN PIPES OF VARIOUS SIZES

Size of Pipe, Inches	Running Foot per Cubic Foot of Contents	Contents in Cubic Feet per 100 Running Feet
3/4	270.00	0.370
1	166.90	0.599
1 1/4	96.25	1.038
1 1/2	70.65	1.415
2	42.36	2.360

TABLE V.—WEIGHTS OF AMMONIA VAPORS AT DIFFERENT GAUGE PRESSURES

Ammonia Gauge Pressure	Weight of 1 Cubic Foot of Vapor, Lb.	Ammonia Gauge Pressure	Weight of 1 Cubic Foot of Vapor, Lb.
0	0.0566	80	0.3304
10	0.0941	90	0.3617
20	0.1269	100	0.3939
30	0.1611	125	0.4766
40	0.1955	150	0.5566
50	0.2292	175	0.6340
60	0.2641	200	0.7188
70	0.2965	...	......

The number of hundreds of running feet of pipe in the system having been determined, the cubic feet contained in it may be found from Table IV. The amount of ammonia necessary may be ascertained by multiplying the cubical contents by the weight of gas per cubic foot corresponding to the pressure to be carried in the pipes when the system is in operation. The weight of ammonia vapor required to fill both high- and low-pressure sides of the system may be determined in this way, in addition to which a liberal margin should be allowed for reserve liquid in the receiver, evaporating liquid in the expansion coils and condensing liquid in the condenser.

Where sharp freezers are in service, a much larger amount of liquid will be required to charge the low-pressure side, the extra charge increasing very rapidly with decreasing pressures.

When the refrigerating machinery is to be operated under average conditions, an ammonia charge figured according to the following tables will be in line with commercial practice.

TABLE VI.—ANHYDROUS AMMONIA REQUIRED FOR THE COMPRESSION SIDE OF REFRIGERATING PLANTS

Tons of Refrigeration	Pounds of Ammonia	Tons of Refrigeration	Pounds of Ammonia
5	110	75	375
10	150	100	440
15	185	150	510
20	230	175	570
25	245	200	620
30	270	225	675
35	290	250	725
40	300	300	840
45	325	400	1040
50	350	500	1215

TABLE VII.—ANHYDROUS AMMONIA REQUIRED PER 100 RUNNING FEET OF PIPE—EXPANSION SIDE

Ammonia for Refrigerating Plants Direct Expansion and Brine Cooling Coils	Size of Pipe	Ammonia for Ice Plants Expansion Coils for Can and Plate Use
14 pounds	1 inch	8 pounds
18 "	1¼ inches	11 "
20 "	1½ "	12 "
25 "	2 "	15 "

In ice plants the amount of expansion surface per ton is more nearly a constant than in direct-expansion refrigerating plants. Since different sizes of expansion piping are used by different builders, the expansion surface does not always bear a fixed relation to the space to be filled with ammonia vapor. The following ammonia charges for ice-making plants may be considered in line with average practice.

TABLE VIII.—AMMONIA REQUIRED FOR ICE-MAKING PLANTS

Tons of ice per 24 hours	5	10	15	25	50	100
Pounds ammonia	100	250	500	1000	2000	4000

The amounts given in Table *VIII* are for the total number of pounds required to charge both high- and low-pressure sides of the ice-making systems in question, while those in Table VI are those required for the compression side only.

Salt

The amount of sodium chloride (NaCl) required to make brine for an ice tank of given capacity is also subject to wide variations on account of the diversity of designs used by the different builders. The factor which most affects the amount of salt necessary is the amount of space left between the bottoms of the cans and the tank. A good general rule is to allow 15 pounds of salt per cubic foot of brine actually required to fill the tank when the cans are in place. Another rough rule is to allow two-thirds ton of salt per ton of ice-making capacity of tank per 24 hours. When calcium chloride (CaCl) is employed, some authorities estimate the amount required at one ton CaCl per ton of ice-making capacity.

CHAPTER VIII

WORKING PRESSURES

It has already been pointed out that a given substance boils at different temperatures under different pressures; the boiling point being raised when the pressure is increased and lowered when it is decreased.* In the case of water, for example, which boils under atmospheric pressure at 212° Fahrenheit, an increase in pressure to 70 pounds gauge raises the boiling point to 316° Fahrenheit, and a reduction in pressure to 29.74 inches vacuum lowers it to 32° Fahrenheit, or to its freezing point. From this, since the law is a general one applying to all known liquefiable gases, it follows that to produce low temperatures the pressure on the refrigerating medium must be reduced to such a point that the corre-

* A common and very simple method of demonstrating that liquids boil at different temperatures as the pressure varies is to partly fill a thin glass flask with water. Apply heat until the water boils vigorously, cork the flask with a rubber cork thru which is inserted a thermometer. Immerse the flask in a vessel of cold water. The condensation of the steam above the liquid in the flask will relieve the pressure and the water will continue to boil even though at a temperature several degrees below 212°. The air having been expelled from the flask by the steam generated before the flask was corked, the removal of a part of the regular atmospheric pressure of 15 pounds per square inch previously exerted on the boiling liquid allows the vapor to pass off more easily, hence the liquid boils at a lower temperature.

Besides increasing the boiling point or temperature at which vapors may separate themselves from the mother liquid, the application of pressure increases the fusing point of all substances that contract when freezing. In both cases the pressure resists the action effected by the application of heat, and additional heat in proportion to the pressure must accordingly be applied to overcome the resistance. In case of substances that contract at the temperature of fusion, such as ice, the application of pressure assists the action. Two pieces of ice pressed firmly together will melt at the point of contact because of the assistance that pressure lends to fusion. Upon the removal of the pressure, however, the pieces will freeze firmly together. This experiment was first performed by Sir Humphrey Davy in 1799 as a means of disproving the then current theory that heat is a substance. Davy's experiment was made in a vacuum at a temperature below the normal melting temperature of ice. Lord Kelvin determined that the freezing point of water is lowered .1235° Fahr. for each atmosphere (14.7 lbs.) increase of pressure.

sponding boiling point will be a sufficient number of degrees below the temperatures to be produced to bring about the heat transfer through the expansion coils or other cooling surfaces. If, for example, it is desired to cool a cold-storage compartment to 10° Fahrenheit, a back pressure of 24 pounds gauge will be found too high to allow ammonia to boil at this temperature. At 23.64 pounds pressure it will boil at exactly 10° Fahrenheit, but since this is the temperature of the surrounding air, there is no difference in temperature to bring about a heat flow and the boiling will not continue. When the pressure is reduced to 19.46 pounds gauge, the ammonia will boil at 5° Fahrenheit, and at this temperature there will be sufficient inflow of heat from the 10° surrounding air to cause quite appreciable refrigeration. A further reduction to 15.67 pounds gauge lowers the temperature of the boiling ammonia to 0° Fahrenheit and the increase in temperature difference from 5° to 10° Fahrenheit will effect a rate of heat transfer just twice as great per square foot of pipe surface as was possible with half the difference. A still further reduction to 9.1 pounds gauge pressure will allow the ammonia to boil at −10° Fahrenheit, at which temperature the heat flow from the 10° room will be twice as great as it was at 15.67 pounds pressure and four times as great as it was at 19.46 pounds. In order to produce the same amount of cooling effect at 19.46 pounds pressure as was obtained at 9.1 pounds pressure, just four times as much pipe surface would have to be employed, and in order to do as much as at 15.67 pounds, just twice the surface would be required.

If, instead of direct expansion, brine circulation is employed, it will be evident that for the same rate of heat flow, a lower pressure and temperature will be required in the latter case. Assuming, for example, that the rate of heat transmission per square foot per degree difference in temperature be the same between the ammonia and brine and between the brine and air as it is between the ammonia and air (an assumption which is not wholly accurate, but which will simplify the example), the heat transmission between the ammonia at 9.1 pounds pressure and the air at 10° Fahrenheit would be only half as great in the case of the brine system as in the case of direct expansion. This because of the fact that 10° difference in temperature must be allowed to cause a heat flow from the air to the brine and another 10° for the flow from the brine to the ammonia. On this basis, in order to produce the same heat flow, 9.1 pounds back pressure would have to be carried in the case of brine circulation against 15.67 pounds in the case of direct expansion. The relationships are expressed in tabular form in Table IX.

TABLE IX.—BACK PRESSURES REQUIRED TO DOUBLE DIFFERENCE IN TEMPERATURE

Back Pressures	23.64	19.46	15.67	9.1 lbs.
Boiling Temperatures	10°F.	5°F.	0°F.	−10°F.
Making a difference in temperature between ammonia and brine of	10°F.; brine and air 10°F.			
And between ammonia and air	20°F.			

PRESSURE AND EFFICIENCY

Where the piping is installed and cannot be increased, there remains only one of the two variables. To increase its cooling capacity, therefore, lower back pressures must be employed.

The saving in first cost by the installation of scanty pipe surface always entails correspondingly lower back pressures, and is soon lost by increased operating expense due to decreased efficiency.

Where there is only one temperature to be produced in the cold-storage compartments, a back pressure is usually carried such that the temperature corresponding to that pressure will be from 10° or less on low-temperature work to 30° or more on high-temperature work, below that of the "cooler" temperature. Under average operating conditions the cost of the amount of expansion pipe required to allow this range in temperatures balances up fairly well with the loss in efficiency that would be encountered if less expansion piping were installed and a lower back pressure carried. The lower the temperature to be produced the lower the efficiency of the machine; consequently a greater expense will be warranted in pipe area so as to increase heat transmission for the smaller temperature range.

Where several different temperatures are to be maintained with one back pressure, no fixed rule can be followed and each individual case must be figured out separately. If only a small percentage of the total cooling be low-temperature work, it is usually advisable to increase the surface and to reduce the temperature range between the liquid ammonia and the surrounding air in the lowest temperature compartment. In this case the use of an abnormal amount of pipe on this small amount of low temperature work tends to increase the efficiency of the whole plant.

While the necessity of producing a low temperature in a single box tends to reduce the efficiency of the entire plant, or that part of it which is required to operate at a low back pressure because of the low temperature, there is a slight compensation for the decreased efficiency in the decreased first cost of expansion piping for the higher temperature boxes. The reduced ammonia pressure occasions a correspondingly reduced ammonia temperature. This increases the range in temperature between the ammonia and the outside substance to be cooled, which in return permits a reduction in pipe areas in proportion to the increase in range.

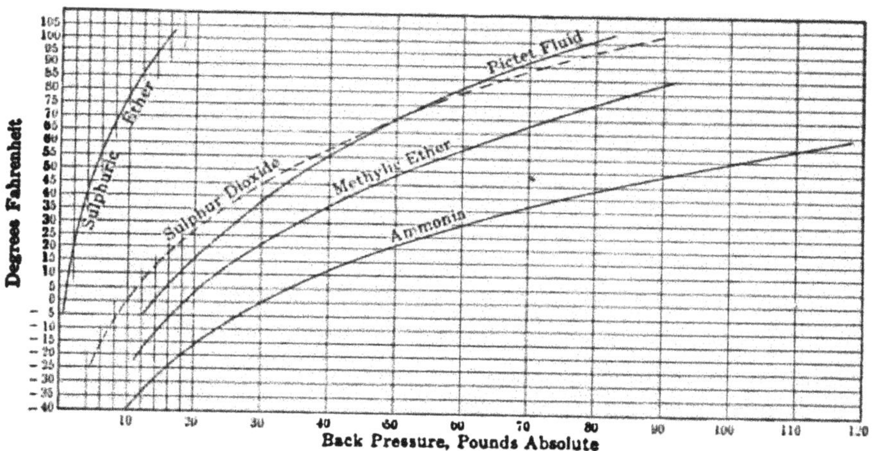

Fig. 39.—Curves Showing Properties of Various Refrigerating Media (Gauge Pressures) = (Absolute Pressures)—(15 Pounds)

Some idea regarding the pressures that should be maintained in expansion coils when operating the different kinds of refrigerating media may be gained by reference to the curves shown in Fig. 39, from which it may be seen that for an expansion temperature of 0° Fahrenheit, such as would ordinarily be employed where temperatures to be produced are from 10° to 20° Fahrenheit, expansion pressures for these different refrigerating media are as follows: Ammonia, 30 pounds absolute; methyl ether, 19 pounds; pictet fluid, 14; sulphur dioxide, 10 pounds.

While no definite rules can be laid down regarding the back pressures that should be carried even under average conditions, the pressures and temperatures given in Table X will be found to be fairly accurate for ammonia.

TABLE X.—BACK PRESSURES AND TEMPERATURES (AMMONIA)

Temperature of room, degrees Fahrenheit	5	10	15	20	28	32	36	40	50	60
Back pressure, pounds gauge	7	10	12	15	22	25	27	30	35	40
Temperature of ammonia, degrees Fahr.	−13	−10	−5	0	8	12	14	17	22	26

In general, the engineer should endeavor to so manipulate his expansion valves as to carry the highest back pressure possible and still produce sufficient refrigeration in his coldest coolers. A second limit to possibilities in this direction is reached when no more ammonia feed can be put on the expansion coils without causing liquid ammonia to return to the compressor, causing it to pound and the rod stuffing boxes to leak. It may be mentioned incidentally that the back pressure can be carried materially higher, and the efficiency of the plant materially increased, by keeping the expansion coils free from the insulating effect of ice on the outside and oil on the inside. The coating of ice should be kept as thin as possible at all times and opportunity should be taken to remove oil from the expansion coils once in every two or three seasons at most, and oftener if much oil is thought to have worked into the system.

When the temperatures get too low, it is better to slow down the machine rather than to reduce the refrigerating effect of the system by closing down expansion valves, reducing the back pressure and literally throwing away power because the refrigeration is not needed. The full importance of this truth is seldom recognized by either the supervising or operating engineer, and it is rarely that either strives for the last pound of increased back pressure half as diligently as for the last inch of vacuum in the steam condensers, although the former pressure is of far more importance than the latter in its effect on the general efficiency of the plant.

Condenser Pressure

Just as the back pressures have to be reduced in order to reduce the boiling point of the refrigerating medium when low temperatures are required, the condenser pressure always rises sufficiently to raise the boiling point of the medium when the temperature of the cooling water is raised. Table XI gives the approximate condenser pressures that should result from the use of different quantities of cooling water of different temperatures on condensers of average proportions.

In every event the condenser pressure should be kept as low as possible and the back pressure as high as the temperatures to be produced will permit, narrow limits between such pressures

TABLE XI.—CONDENSER PRESSURES AND TEMPERATURES (AMMONIA)

1 gallon per minute per ton per 24 hours—							
Temperature of cooling water, degrees Fahr.	60	65	70	75	80	85	90
Condenser pressure, pounds gauge	183	200	220	235	255	280	300
Temperature of condensed liquid ammonia, degrees Fahrenheit	95	100	105	110	115	120	125
2 gallons per minute per ton per 24 hours—							
Condenser pressure, pounds gauge	130	153	168	183	200	220	235
Temperature of condensed liquid ammonia, degrees Fahrenheit	77	85	90	93	100	105	110
3 gallons per minute per ton per 24 hours—							
Condenser pressure, pounds gauge	125	140	155	170	185	200	215
Temperature of condensed liquid ammonia, degrees Fahrenheit	75	85	90	93	95	100	105

Ammonia condenser pressures resulting from the use of different quantities of cooling water at different temperatures.

being as important to the efficiency of a refrigerating system as wide ones are to that of a steam engine, in which the economy increases with the range between boiler pressure and condenser pressure.

Freezing Back

It often happens that the expansion valves are not properly adjusted, or that the expansion coils are so arranged that, like poorly designed boilers, there is abnormal entrainment and considerable liquid ammonia is carried back with the returning vapor. In this case the scale separator may act as a veritable separator and temporarily interrupt the passage of the entrained liquid. On account of the difficulty of returning any liquid so trapped to the expansion coils the scale traps are of little value as separators except as means of keeping occasional large volumes of liquid from returning to the compressor. Once having become filled with liquid ammonia they remain in this condition for some time. Since in order to evaporate the ammonia must have heat, and since the temperature of the boiling ammonia corresponding to the back pressure usually carried in refrigerating and ice-making work is sufficiently low to produce ice on the outside of the traps, piping, etc., these parts soon become heavily insulated with ice which further materially reduces the amount of heat that can be absorbed, and the entrained liquid enters the compressor with a considerable capacity for absorbing heat. If this amount is abnormal it may cause the compressor to pound for the same obvious reason that a steam engine pounds when it receives a quantity of entrained water in the steam. When such abnormal quantities of liquid enter the compressor cylinder, it is usually evidenced by the abnormal cooling effect on the compressor walls, or more noticeably that of the piston rods which may contract sufficiently to allow the ammonia to leak by the packing. The evaporation of this entrained liquid ammonia in the compressor cylinder, or that introduced directly into the cylinder through an expansion valve designed for that purpose, refrigerates the gas as well as the compresssor parts and tends to prevent superheating of the gas during compression.

The evaporation of the liquid ammonia remaining in the expansion coils when the compressor is shut down causes the rise in back pressure usually so noticeable a few hours after the plant has been shut down.

Condition of Ammonia

The condition of the ammonia vapor as regards saturation or supersaturation may best be arrived at through thermometers inserted in mercury wells in the return and discharge lines near the compressor. Tables of "properties of saturated ammonia" indicate at a glance the temperatures at which the vapors should return to the machine under different conditions of back pressure and assumed saturation.

If the last trace of the liquid ammonia is evaporated before the vapors reach the compressor, and the return pipes are uninsulated, there is likely to be considerable superheating, i.e., the temperature of the vapor entering the compressor is likely to be several degrees higher than that shown by the tables of properties of saturated ammonia gas to correspond to the back pressure carried. This condition results in a considerable loss of efficiency and should not be allowed to continue.

While difference in opinion regarding the amount of unevaporated liquid the return ammonia gas should contain in order to give maximum efficiency has given rise to two distinct systems, viz., the "wet" and the "dry" compression, a discussion of the relative merits of the two systems would be too far-reaching to warrant its introduction here. The best general rule regarding the wet or dry operation of compressors is to follow the instructions of the respective builders.

THE FROST LINE

In the absence of more accurate means, such as thermometers, for determining the temperature of the returning ammonia gas, the "frost line" has been forced into service to give at least some slight indication of such temperatures. The simple formation of frost on the outside of a pipe containing cold ammonia gas, or, in fact, any other cold medium, indicates nothing more nor less, however, than that the heat from the outside atmosphere is absorbed with sufficient rapidity to reduce the temperature of the pipe and nearby air to at least 32° Fahrenheit, under which condition atmospheric moisture is first precipitated, just as rain or dew is formed when moisture-laden air becomes cooled by heat radiation to air at a lower temperature or contact with other colder objects, and, second, is frozen, just as dew is frozen to form frost when its temperature is reduced below 32° Fahrenheit.

If there is liquid ammonia enough at the expansion valve, frost can be carried the full length of a coil of almost any length and clear back to the machine, if desired, at a back pressure of 25 pounds, because the temperature of saturated gas at 25 pounds pressure is 11.5° Fahrenheit, which is 20.5° Fahrenheit below 32°, the freezing point of water. That a coil does not frost to the end under a back pressure of 25 pounds indicates that either there is an insufficient supply of liquid ammonia at the expansion valve or that there is an obstruction which prevents a sufficient amount from passing the expansion valve. An obstructed expansion valve is indicated by there being little or no change in the sound of the passing liquid when the valve is opened several turns. Such obstructions can often be removed by the sudden opening and closing of the expansion valve.

Since the formation of frost on an ammonia pipe is influenced by the room temperature, it cannot be an ideal means of judging temperatures within the pipes. Where considerable entrained liquid ammonia is present, the general appearance of the frost formed, or the way one's wet finger sticks to the pipe, may give some slight indication of the action taking place inside. Where low temperatures are carried, the return gas may be so far below 32° Fahrenheit that the same rise in temperature that would ordinarily completely change the appearance of the return line, if it took place at a higher temperature, would not affect the appearance of the frost line at all.

Insulated Suction Lines

It may be generally asserted that expenditure of energy is necessary to remove heat from any substance at any temperature to another substance at a higher temperature. If, then, a certain amount of the heat in the returning ammonia gas has its origin in the engine room, where its absorption is manifested by frost on the return line to the compressor, it is evident that additional energy will have to be expended in the engine that drives the compressor, which energy costs coal, labor and, finally, money. The return lines to compressors should be effectively insulated to reduce this loss. Nothing is more erroneous than the argument that because the returning gas has passed the rooms that it is sent out to cool, there will be no loss because of the heat absorption through exposed, uncovered cold pipes. The useless expenditure of a single unit of refrigeration is just as prodigal as the throwing away of an equivalent amount of money. The fact that such losses are allowed to continue in some of the largest refrigerating and ice-making plants in the country is poor excuse for their existence in others.

CHAPTER IX

CLEANING THE SYSTEM

DEFROSTING REFRIGERATOR COILS

The problem of defrosting cold storage coils is one which is too often ignored at the time of making the installation. Where such has been the case, the operator often finds himself working at great disadvantage, not only because of his inability to produce the required temperature, but also because of the decreased efficiency of the mechanical equipment entailed by the reduction in back pressures necessary to coax the heat from the cold-storage rooms through the additional resistance offered to its passage by the accumulated ice. This disadvantage affects principally the coal bill and fortunately for the operator, though unfortunately for the owner, is in the majority of cases not recognized, or, if apprehended, is not charged to the proper account.

The effect of a coating of ice on direct-expansion pipes may be shown as follows: Assuming a heat transfer of 10 B.t.u., in round numbers, per hour, per square foot per degree of difference in temperature inside and out, for a flat metallic refrigerating surface,* and an equal amount for a sheet of ice one inch thick, it follows that the heat transmission through a square foot of direct-expansion cooling surface insulated with a layer of ice one inch thick will be only one-half that of the uncoated surface. As a matter of fact, it would seem from the context that the value of 10 B.t.u. given as the heat conductivity of ice applies to plate-ice conditions under which the wetted surface of the submerged ice will transmit materially more heat than a dry surface in contact with air. This would indicate that the decrease in heat-transmitting capacity of direct-expansion surfaces in air due to a coating of ice is even more than 50 per cent. This condition will be partially offset by the fact that on account of the increasing diameter the layer of ice in the case of cylindrical surfaces such as pipes, which, together with the fact that such coatings usually present an irregular surface, further increasing the heat-absorbing area,

* See Siebel's "Compend of Mechanical Refrigeration and Engineering," pages 190 and 207.

may increase the heat transmission sufficiently to make up for the lesser heat transfer between the air and dry ice and make 50 per cent at least a reasonable estimate of the loss in heat-absorbing capacity due to one inch of ice.*

BRINE COILS

Brine pipes may be readily defrosted by the circulation of hot brine. This may be accomplished through the main feed and return headers where the operation does not have to be performed very frequently, or as in abattoirs where the excessive amounts of moisture from the hot meats to be chilled make the accumulation of frost very rapid, or by a separate set of defrosting headers.

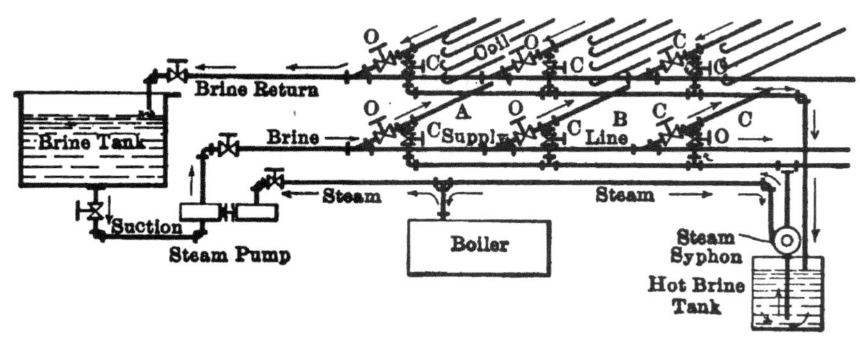

Fig. 40.—Diagram of System for Defrosting Brine Coils

An arrangement such as is illustrated diagrammatically in Fig. 40 allows a single coil to be defrosted without interfering with the operation of the remaining coils, which is almost an absolute necessity where temperatures are to be kept within control in abattoir chill rooms. The heating of the brine may be effected in a small auxiliary tank by the exhaust steam from the pump used for circulating the hot brine, or it may be done by the action of a small steam siphon.

The branches of the main supply and return brine headers are provided with small pipe connections on the coil sides of the coil supply and return valves. The connections, each provided with a valve or stopcock, communicate into small supply and return headers, the former connecting with the discharge side of the steam siphon and the latter discharging into a tank or barrel into

* Under average commercial conditions of intermittent frosting a square foot of direct-expansion surface in air is usually credited with a heat transmission of only from 2 to 4 B.t.u. per hour per degree difference in temperature.

which the siphon suction line runs. By closing the main supply and return valves on any one or any number of the coils to be defrosted, and opening the corresponding valves connecting with the defrosting headers, a direct circuit is established from the siphon through the coil and back to the barrel. The admission of the steam to the siphon starts the circulation and at the same time, since the action of the siphon depends on the condensation of the steam in the brine circulated, the brine is gradually heated. That this heating should take place slowly is of importance in the successful defrosting of long coils, especially in the case of the lock seam and spiral riveted galvanized iron pipes that have been so commonly used.

Direct-expansion Coils

In the case of direct-expansion coils, the defrosting method probably most satisfactory where the cold-storage temperatures are above 32° Fahrenheit is to install sufficient coil surface to allow a part of the coils to be shut off at any time, so that the frost will melt without artificial heat and at the same time produce a certain amount of useful refrigeration. If it is necessary to force the defrosting process by the use of outside heat, a hot gas line from the condenser may be connected to the liquid-line connections to the separate coils just inside the expansion valves. The hot gas, after melting the ice as it passes through the coils, returns to the compressor together with the return gas from the remaining coils.

Where the temperatures carried in the cold-storage compartments are below 32° Fahrenheit, and in which the defrosting cannot be effected without the use of artificial heat, often very objectionable, two methods are available, viz.: that of forcibly removing the ice with scrapers and that of suspending over the pipes trays of calcium chloride. This substance is exceedingly deliquescent salt, which in absorbing moisture from the air forms a saturated calcium brine which freezes at a very low temperature. In trickling down over the coils, the brine melts the ice, forming a more dilute brine, which is then conducted away to the sewer, or, if the quantities involved warrant the expenditure of labor, may be evaporated and the calcium chloride recovered.

Oil in the Refrigerating System

Next to unintelligent design, which sometimes provides for the operating engineer a plant that cannot be made to develop nearly as high efficiency as operating conditions should warrant, and unintelligent operation, which fails to get nearly as high efficiency as operating conditions and mechanical design should warrant, the worst foe to economy is foreign matter in the refrigerating system.

In order that the oils used in the system shall not stiffen prohibitively at the low temperatures encountered and not be saponified by the ammonia, only very light mineral oils can be employed. Such oils range from 22° to 30° Baume, corresponding to a specific gravity of from 0.924 to 0.88. These oils should have a cold test of about 0° Fahrenheit, to obtain which they will have a flash point of between 310° and 400° Fahrenheit. This low flash point implies that a considerable amount of vapor will be given off at a much lower temperature. Since discharge temperatures of compression machines often approach these temperatures, it is obvious that a considerable amount of oil will pass to the condenser, not as a liquid but as a vapor. Under such conditions, since there is no material cooling effect in the oil separator, only liquid oil would be precipitated at that point.

Baffle plates on which the discharged gas impinges and to which particles of oil will tend to stick by adhesion as well as on account of the abrupt reversal of direction of flow of the gas, which tends to separate out the slightly denser substances by centrifugal force, are precautions in the right direction; but except when discharge temperatures are low it is useless to expect to intercept all of the oil until it has been actually condensed. This operation being carried out most effectively in the condenser, the oil will pass to the receiver with the liquid ammonia, where, since there is a difference in specific gravity of 0.23 to 0.27, the separation can be more easily effected

Oil in Coils

The question of removing oil from the expansion coils, whether they are used for chilling cold-storage rooms, for chilling brine to be circulated through cold-storage rooms or for chilling the brine of an ice tank, is one which has received much attention, but, although much experimenting has been done and a few patents have been issued, no cheap and effective method has yet been devised. Where coils will drain so that there will be no danger of entrapping condensed moisture, the most effective method of removing the oil is to pump out the coils, disconnect them and blow them out with the highest pressure live steam available, letting the steam blow through each coil as long as any oil appears at the exhaust end. The brine may be left in the tank and heated up by steam, so as to prevent, as far as possible, the condensation of steam inside the pipes, or if the brine has been removed the same result may be obtained by leaving the covers on the tank and blowing in live steam.

After the steam has been blown through the coils until no more oil appears, each coil should again be blown out with air to remove any traces of moisture from the steam. The moisture-absorbing capacity of the air can also be increased by heating it before it is introduced into the coils. The coils themselves may be heated as previously described. Long coils cannot be perfectly cleaned cold by air alone. Where coils will not drain in the tank as installed, the only effective way of extracting the oil is to remove the brine and disconnect them at such points as will enable them to drain, or remove them from the tank entirely. In case this method cannot be employed, ammonia purifiers may be installed with advantageous results. These purifiers work continuously when the plant is in operation, and their action can be facilitated by frequently pumping out the system. While it can hardly be expected that all of the oil can ever be removed by this means, their slow continuous action is of great benefit in plants in which it is impossible to shut down in order to employ more effective methods, and will largely prevent the accumulation of oil in new or recently cleaned systems.

Permanent Gases

The action of so-called permanent gases in the condenser is less detrimental than that of oil in the expansion coils which forms an insulating lining of fairly high efficiency on the inside of the heat-absorbing surfaces. The principal effect of these gases, if their operation can be limited to the condensers, is to occupy space that should be available for the ammonia gas. This reduces the effective area of the condenser cooling surface, causing an increased head pressure with the usual amount of cooling water or an increased amount of cooling water to maintain the same head pressure.

These so-called permanent gases are of rather uncertain origin and, as the name implies, are gases not liquefiable under the same conditions of temperature and pressure as ammonia. In a new plant or one recently overhauled they may consist almost wholly of atmospheric air. In an old one they are more likely to consist of nitrogen and hydrogen from decomposed ammonia, vapors of oil and mixtures of other hydrocarbon gases formed by the action of heat on hydrogen in the presence of vapors of oil, and water and other impurities contained in ammonia.

There is a considerable difference in opinion as to whether or not these gases are heavier or lighter than ammonia, and accordingly whether they can be most readily purged from the highest or the lowest part of the system. It seems reasonable to suppose that on account of the comparatively rapid flow of the ammonia vapors through the system, that sooner or later these gases, wherever formed, will find their way to the condenser. If the liquid outlets from the condenser are properly sealed, there will be no other way for them to escape than through the purge valve at the top of the condenser. If these gases are lighter than ammonia, they will accumulate at the top of the condenser, and if heavier they will be driven to the top by the reëxpanding ammonia in the bottom of the condenser as soon as the pressure is removed by the opening of the purge valve.

Purging

Condensers should be purged as often as the accumulation of gases indicate that they need it. To do this most advantageously, the liquid outlet and hot-gas inlet valves of the coils to be purged should be closed and a liberal supply of cooling water allowed to flow over them for some time. The permanent gases can then be purged out through a small rubber tube, one end of which is connected to the purge valve; the other end should be immersed in a pail of water. If permanent gases escape when the purge valve is slowly opened, bubbles will rise to the surface of the water. If only ammonia escapes, the bubbles of ammonia gas will be dissolved in the water before reaching the surface, giving rise to a sharp crackling sound such as that caused by the condensation of steam in the process of heating water by the direct admission of steam through an open pipe.

When the water in the pail has become saturated and accordingly cannot absorb more ammonia, the bubbles will rise to the surface and it is difficult to discriminate between ammonia and the other gases. For this reason it is advisable to replace the aqua ammonia formed by fresh water often enough so that it will not become saturated.

This requires that in addition to keeping the inside of the condensers free from obstructing gases, economy of power and cooling water further require that the outside of the condensers be kept as free as possible from incrustations left behind by the cooling water. Scale, like ice on the outside and oil on the inside of the expansion coils, has the effect of insulating the heat transmitting surfaces, causing either higher condenser pressure or requiring the use of an excessive amount of cooling water.

Incrustation on Condenser Coils

While the comparatively high working temperature of condenser coils, together with the usually ample provisions for draining each separate coil, prevents the accumulation of such large quantities of oil as are often lodged in expansion coils, condenser coils are exposed to another source of loss of efficiency from without. Where the available cooling water is abnormally hard, or carries a large amount of suspended matter, ammonia condensers, and especially steam condensers, soon become coated with a deposit of scale or mud, which, if not properly removed, becomes a more or less effective insulator according to the composition of the deposit. The heat conductivity of metallic surfaces is not the same per degree difference in temperature at medium and low as it is for high temperatures, and it does not therefore follow that the resistance offered by the scale accumulating on the outside of atmospheric and submerged ammonia and steam condensers is the same as that of scale on the inside of a boiler. However, some slight idea of the extent of the loss may be gained from the fact that in steam-boiler practice, the insulating effect of scale results in a thermal loss corresponding to about 2 per cent of the fuel for each 1-64 inch in thickness of scale.

It seems superfluous to state that the heat-absorbing surfaces of brine-cooling coils and the heat-radiating surfaces of condenser coils should always be kept covered with brine and condenser water respectively. Nevertheless, it is not uncommon to see refrigerating plants operating with the brine so low in the tanks that the top expansion coils are exposed and the distribution of water over the condensers so irregular that a large portion of the surface is dry.

CHAPTER X

CAPACITY OF REFRIGERATING MACHINES

In general there are two methods by which to determine the amount of cooling effect produced by a refrigerating machine.

Effect on Secondary Refrigerating Mediums

The first method is to measure the results produced as, for example, the cooling effect in degrees produced on a known quantity of brine or other mediums having known specific heats. The number of units of heat extracted per minute, hour or day being known, the capacity of the machine in pounds or tons of equivalent ice-melting capacity may be readily computed by dividing the number of units of heat extracted by the number of such units required in the same length of time to produce refrigeration at the rate of one ton per 24 hours.

Effect on Primary Refrigerating Mediums

The second method is to actually weigh or compute the number of pounds of the primary refrigerating fluid, assumed in this case to be ammonia, passing through the cycle of operations per unit of time and, by knowledge of the amount of refrigerating effect that a pound of the refrigerating fluid is capable of producing under the observed conditions of back pressure and liquid temperature,* finally arrive at a more or less accurate estimate of the amount of cooling effect being produced.

Units

The units involved in making such calculations are as follows:

(1) *British Thermal Unit* (*B.t.u.*).—Equivalent to the specific heat of water, or the amount of heat required to raise the temperature of a pound of water through 1° Fahrenheit at its temperature of maximum density, 39° Fahrenheit.

(2) *Pound of Refrigeration.*—Equivalent to the expenditure of negative heat (absorption of heat) at the rate of 144 B.t.u. per

* On account of the usual inaccuracy of pressure gauges, accurate determinations of liquid temperatures can best be made by means of thermometers set in mercury wells inserted in the liquid lines just before they enter the expansion valves.

twenty-four hours; 144 B.t.u. being the latent heat of ice, or the amount of heat required to melt a pound of ice "from and at" 32° Fahrenheit (ice at 32° melting into water at the same temperature).

(3) *Ton of Refrigeration.*—Equivalent to the expenditure of negative heat (absorption of heat) at the rate of 2,000×144 B.t.u., or 288,000 B.t.u. per twenty-four hours; 288,000 B.t.u. being the amount of heat required to melt a ton of ice "from and at" 32° Fahrenheit.

From the foregoing it is apparent that the capacity of a refrigerant, which extracts heat at the rate of 288,000 B.t.u. per twenty-four hours, is one ton. The above rate is equivalent to 288,000÷24=12,000 B.t.u. per hour, or 288,000÷(24×60)=200 B.t.u. per minute.

To absorb heat at this rate certain definite quantities of the primary refrigerating fluid must be evaporated per minute, hour or day; in addition to which if a secondary refrigerating medium, such as brine, is employed, a certain quantity must be cooled through a sufficient range of temperature to supply the amount of heat required to evaporate the primary fluid.

Standard Conditions

It has been proposed to employ as standard 185 pounds head pressure—under which pressure condensed anhydrous ammonia leaves the condenser at about 90° Fahrenheit—and 15.67 pounds back pressure, corresponding to an evaporating temperature in the cooler of 0° Fahrenheit. Under these standard conditions, where the anhydrous ammonia enters the refrigerator at 90° Fahrenheit and evaporates at 0° Fahrenheit, from 27 to 28 pounds of liquid must be evaporated per hour per ton of refrigerating effect produced. This means that the ammonia compressor must have an effective displacement capacity of about four cubic feet per minute per ton.

Effect of Pressure

The refrigerating capacity of evaporating refrigerants does not depend on the volume of gas evolved, except as volume depends on weight. The volume of gas varies widely with the pressure, but aside from the cooling effect that must be expended on the liquid to cool it from the condenser temperature down to the cooler temperature, the weight of refrigerant per unit of cooling capacity produced is practically constant under all conditions of temperature and pressure.*

At 10 pounds back pressure, for example, corresponding to an evaporating temperature of about 8° below zero, the volume of ammonia gas is 10.8 cubic feet per pound. At 32 pounds back pressure, corresponding to an evaporating temperature of about 19°, the volume is about 6 cubic feet. This means that operating at 10 pounds back pressure a compressor must pass approximately 66 per cent more volume of gas per unit of capacity than when operating at 32 pounds pressure.

Computed Capacity—Example

The method of arriving at the size of the compressor required for the performance of a given refrigerating duty per twenty-four hours is as follows: The latent heat, or the amount of heat expressed in B.t.u., required to evaporate a pound of anhydrous ammonia at 10 pounds back pressure is about 560. If, for example, 150 pounds condenser pressure is carried, the liquid ammonia will pass to the refrigerator at about 84 degrees. The evaporating temperature corresponding to 10 pounds back pressure is −8°, making a difference of temperature of 92° through which the liquid must be cooled before it can produce any useful cooling effect in the refrigerator. For approximations, the specific heat of the ammonia liquid may be taken to be the same as that of water, i.e., unity, in which case the expenditure of 92 B.t.u. will be necessary to cool the liquid from the temperature of the condenser to that of the refrigerator. Subtracting this amount from the latent heat—560 B.t.u.—we have 468 B.t.u. remaining for the performance of useful work in the refrigerator.

Pounds Refrigeration

The latent heat of ice, taken as the unit pound capacity of refrigeration, is 144. The evaporation of one pound of anhydrous ammonia under the foregoing conditions will produce $468 \div 144 = 3.25$ pounds of refrigeration; this regardless of the time in which the evaporation takes place.

Pounds of Ammonia

The evaporation of one pound of anhydrous ammonia per min-

* This presupposes only limited superheating when the refrigerant in question is in the gaseous state.

ute under these conditions would produce refrigeration at the rate of 468 ÷ 200 = 2.34 tons per twenty-four hours. At the rate of one pound per hour the tonnage rate per twenty-four hours would be 468 ÷ 12,000 = 0.039.

Cubic Feet of Ammonia

At 10 pounds back pressure each pound of anhydrous ammonia evaporated has a volume of about 10.8 cubic feet. A compressor displacing this volume of gas per minute will have a capacity of 2.34 tons, or a ton capacity for every 10.8 ÷ 2.34 = 4.6 cubic feet of piston displacement per minute.*

Capacity of Compressor

The method of computing the capacity of a refrigerating machine of a given size when operating at a given number of revolutions, from the apparent displacement in cubic feet per minute, involves two other very important factors: First, the back pressure at which the compressor is operated; second, the displacement efficiency of the particular compressor in question when operating at the back pressure in question.

The amount of refrigeration produced is directly dependent on the number of pounds of the refrigerating fluid evaporated, due allowance being made for the range of temperature through which the liquid must be cooled before it can do useful work in the refrigerator, and the cooling effect available from superheating of the gas.

Displacement Efficiency of Compressor

When the compressor itself is employed as a meter, i.e., when the amount of the refrigerating medium is computed from the number of cubic feet of volume swept out per minute by the piston, it is necessary to assume or determine the compressor displacement efficiency in order to arrive at the *actual* displacement from the *apparent* displacement indicated by the volume swept out. The weight of a refrigerating medium vapor is directly proportional to its absolute pressure. For a given back pressure the weight of the gas per cubic foot can be determined directly from published tables of the properties of the refrigerating medium in question.

* According to the standards adopted by the Ice Machine Builders in 1903, the evaporation of 27.7 pounds of ammonia under the "standard" conditions of 185 pounds head pressure (90° Fahrenheit) and 15.67 pounds back pressure (0° Fahrenheit) constituted a ton capacity. On this basis, approximately 5 cubic feet of displacement would be allowed per ton per twenty-four hours, in a compressor of 80 per cent. displacement efficiency.

The displacement efficiency of compressors, or the ratio of their apparent to their actual cubical displacement, is not easy to determine, and it is the exception rather than the rule that the builders of compressors can themselves give the efficiency of their various machines when operated under different heads and back pressures. As a matter of fact, the only exact means of determining such efficiencies is by the laborious method of checking the amount of refrigerating fluid apparently passed through the compressor, with the amount, determined by weight, actually passed through the expansion valve. Even this method has a serious shortcoming in the case of "wet" compression machines in that a considerable amount of unevaporated liquid may pass through the compressor and this liquid appearing in the weights of liquid passed through the expansion valve indicates a higher displacement efficiency than the compressor deserves.

Apparent Cubical Displacement

To compute the refrigerating capacity of an ammonia compressor of the assumed cylinder dimensions of 19x38 operating at 45 revolutions per minute under the above suggested standard conditions of 185 pounds head and 15.67 pounds back pressure:

If r = radius, or one-half cylinder diameter,
l = length of stroke, both in inches,
n = number of revolutions per minute.

Then A = area of a 19-inch piston =

[1] $A = \pi r^2$ or $3.1416 \times 9.5 \times 9.5 = 283.53$ sq. in;
V = volume of 19x38-inch cylinder.

[2] $V = \dfrac{\pi r^2 l}{1728} = \dfrac{283.53 \times 38}{1728} = 6.235$ cu. ft.

D = apparent displacement per minute double-acting compressor.

[3] $D = 2\dfrac{\pi r^2 l n}{1728} = 2 \times 6.235 \times 45 = 561.15$ cu. ft.

Somewhat simplified this expression becomes

[4] $D = d^2 l n\ 0.00090903$; or, since

$\dfrac{l n}{6} = \dfrac{2 l n}{12}$ = piston speed $P S$ in feet per minute.

[5] $D = d^2 P S\ 0.00545418$ and $19 \times 19 \times 885 \times 0.00545418 =$ 561.15 cu. ft.

Effect of Working Pressures

The amount of heat absorbed in the evaporation of any liquid depends on the temperature of the liquid supplied and the pressure under which it evaporated as well as on the latent heat of vaporization of the liquid. Water, for example, fed to a boiler at a temperature below that of the steam generated must be heated up to the boiling point before it can be evaporated. Liquid ammonia fed into expansion coils at a temperature above that of the ammonia vapors must similarly be cooled down to the boiling point corresponding to the pressure. Since the only means of cooling the latter is by its own evaporation, it is evident that just so much liquid as evaporates in cooling itself can do no useful refrigerating work on other products. The greater the range in temperature through which it must be cooled, whether on account of the abnormally high liquid temperature or low evaporating temperature, the less will be the useful cooling effect available per pound of refrigerant.

The total heat absorbing capacity, $R\,b\,p$ of a pound of liquid refrigerant, is known as its latent heat of vaporization R. And its value depends on the back pressure $b\,p$ at which it evaporates. The amount of heat Q that must be extracted to cool a pound of liquid refrigerant from the temperature corresponding to the head pressure $h\,p$ down to that of the back pressure $b\,p$ is the difference between the values of the "heat of the liquid" Q under the two conditions. The available cooling effect C per pound of refrigerant is accordingly

$$[6]\quad \left\{\begin{array}{c} C \\ \text{Available} \\ \text{Cooling ef-} \\ \text{fect per} \\ \text{pound} \end{array}\right\} = \left\{\begin{array}{c} R\,b\,p \\ \text{Latent heat} \\ \text{of vaporiza-} \\ \text{tion at back} \\ \text{pressure } b\,p \end{array}\right\} - \left[\left\{\begin{array}{c} Q\,h\,p \\ \text{Heat of the} \\ \text{liquid under} \\ \text{head pres-} \\ \text{sure} \end{array}\right\} - \left\{\begin{array}{c} Q\,b\,p \\ \text{Heat of the} \\ \text{liquid under} \\ \text{back pres-} \\ \text{sure} \end{array}\right\}\right]$$

As tables of the properties of refrigerating mediums do not always give values for the "heat of the liquid" at different temperatures, the number of heat units required to cool the liquid may be arrived at by multiplying the number of degrees $(t_1 - t_2)$ through which the liquid is cooled by S, the specific heat of the liquid, and expression [6] becomes [7], in which have also been substituted values corresponding to standard conditions.

Since the displacement of ammonia compressors is expressed in cubic feet, the available cooling effect per cubic foot of gaseous

$$[7] \left\{\begin{matrix} C \\ \text{Available} \\ \text{cooling} \\ \text{effect per} \\ \text{pound—} \\ 465.5 \end{matrix}\right\} = \left\{\begin{matrix} R\,b\,p \\ \text{Latent} \\ \text{heat of} \\ \text{evapo-} \\ \text{ration at} \\ \text{back} \\ \text{pressure} \\ (15.67) \\ 555.5 \end{matrix}\right\} - \left\{\begin{matrix} S \\ \text{Specific} \\ \text{heat} \\ \text{taken as} \\ \text{unity} \\ * \\ 1 \end{matrix}\right\} \left\| \left\{\begin{matrix} t_1 \\ \text{Temper-} \\ \text{ature} \\ \text{corre-} \\ \text{sponding} \\ \text{to head} \\ \text{pressure} \\ h\,p \\ 90^\circ \end{matrix}\right\} - \left\{\begin{matrix} t_2 \\ \text{temper-} \\ \text{ature} \\ \text{corre-} \\ \text{sponding} \\ \text{to back} \\ \text{pressure} \\ b\,p \\ 0^\circ \end{matrix}\right\} \right\|$$

B.T.U. PER CUBIC FOOT

refrigerant passing through the compressor is most frequently employed. This may be readily determined by dividing the expression [7] by V, the volume occupied by a pound of ammonia under back pressure $b\,p$ and for standard conditions becomes

$$[8]\quad C = \frac{R\,b\,p \times S\,(t_1 - t_2)}{V} = \frac{555.5 - 1\,(90 - 0)}{9.028} = 51.56 \text{ B.t.u.}$$

If the apparent displacement in cubic feet per minute as determined by expression [5] be multiplied by the cooling effect per cubic foot as determined by expression [8], the result will be the apparent capacity of the compressor expressed in B.t.u. per minute, which value divided by 200 gives the apparent tonnage capacity of the compressor per twenty-four hours. Expressed as an equation:

$$\left\{\begin{matrix} \text{Cooling} \\ \text{effect in} \\ \text{tons per} \\ \text{24 hrs.} \end{matrix}\right\} = \frac{\left(\begin{matrix} D \\ \text{Apparent} \\ \text{displacement} \\ \text{per min.} \end{matrix}\right) \times \left(\begin{matrix} C \\ \text{Cooling effect} \\ \text{in B. t. u. per} \\ \text{cu. ft.} \end{matrix}\right)}{\left(\begin{matrix} \text{B. t. u. per min.} \\ \text{equivalent to one} \\ \text{ton per 24 hrs.} \end{matrix}\right)}$$

$$= \frac{561.15 \times 51.56}{200} = 144.66 \text{ tons.}$$

CUBIC FEET DISPLACEMENT PER TON

Since 200 B.t.u. per minute is the equivalent of a ton per twenty-four hours, 200 divided by 51.56, the number of B.t.u. of cooling effect available per cubic foot under standard conditions

* For various values of the specific heat of anhydrous ammonia determined by a number of authorities see Transactions of A. S. M. E., pp. 522–3, Vol. 26, 1905.

gives 3.88, the number of cubic feet of gas that must be actually displaced per ton. Assuming that the compressor has a displacement efficiency of 80 per cent the apparent displacement per ton will be $3.88 \div .80 = 4.85$ cubic feet.*

APPROXIMATE NOMINAL CAPACITY OF COMPRESSORS

Dividing the constants of equations [4] and [5] by 5 (see footnote) gives equations [9] and [10], which will be found convenient in arriving at the approximate nominal capacity of a compressor when its dimensions and speed of operation are known.

[9] $\text{Tons} = d^2\, l\, n\ 0.000181806$

[10] $\text{Tons} = d^2\, P\, S\ 0.00109081$

Substituting values as above,

$$= 19 \times 19 \times 38 \times 45 \times 0.000181806 = 112.23 \text{ tons.}$$

$$= 19 \times 19 \times 285 \times 0.00109081 = 112.23 \text{ tons.}$$

The above equations give approximate results only, and only in the special case of standard conditions and approximately 80 per cent compressor displacement efficiency. To find the capacity of a compressor in the general case in which it is operated at other back and head pressures, the cooling effect per cubic foot actual displacement must be determined in each case by equation [8].

Furthermore, since the efficiency of ammonia compressors is subect to wide variations, both through diversity of design and diversity of operating conditions to which the same machine is often subjected, the efficiency of the individual compressor should be determined under its own operating conditions. This can be accomplished most accurately by determining the quantity of refrigerating fluid actually passing through the system and comparing this amount with the apparent amount computed from the displacement of the compressor.

CAPACITY DETERMINED BY TEST—WEIGHING PRIMARY REFRIGERANT

The best way to determine the amount of liquid refrigerant is to weigh it. Fig. 41 represents a condenser, a pair of weighing tanks and their connections. For testing, crosses are inserted in

* The above is arrived at on the basis of unity taken as the latent heat of anhydrous ammonia. The somewhat higher value of 1.1 sometimes employed would make the required actual displacement about 4 cubic feet and the apparent about 5.

the inlet and outlet lines, and valves and additional pipes are attached as indicated. When using the weighing tanks, the outlet valve *D* is closed or blanked off and, as the valves in the new connections are closed, the liquid refrigerant collects in the receiver.

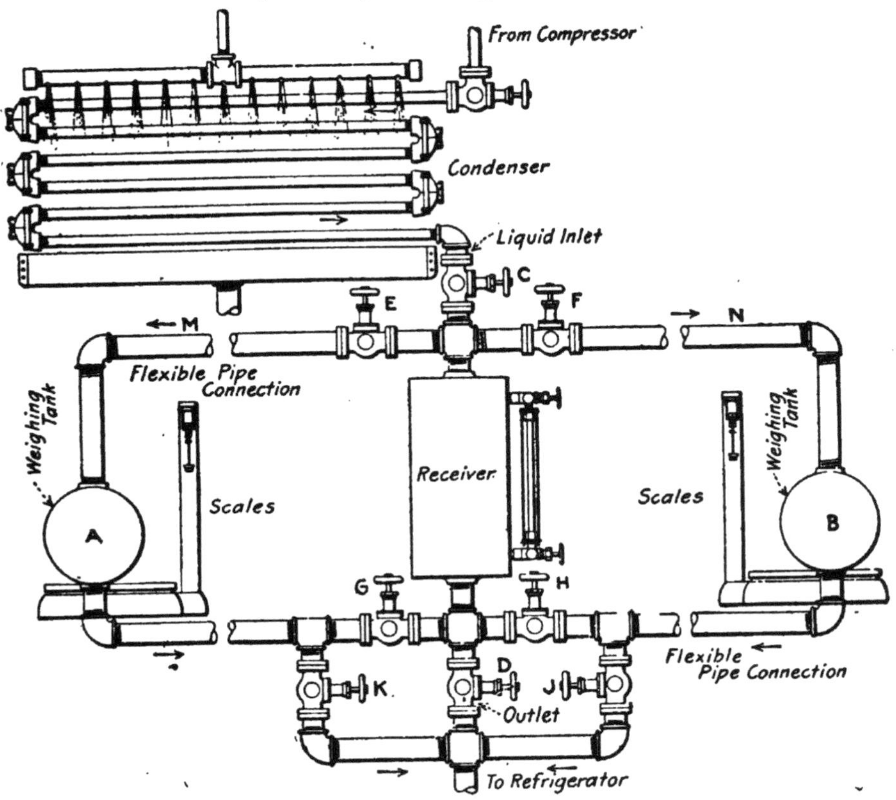

Fig. 41.—Diagram of System for Weighing Liquid Refrigerant

The weighing tank *A* is filled by opening valves *E* and *G*, after which valve *G* is closed and the gross weight of the tank and its contents is determined. The weight of the refrigerant is then found by subtracting the net weight of the tank and the liquid in the bottom connections. While the liquid in tank *A* is being weighed, tank *B* is supplying the cooler through valve *J*. Alternate filling and emptying of the two tanks allows the operation of the plant to proceed without interruption. When employing this method for weighing the refrigerating liquid, it is necessary that the pipes connecting with the weighing tanks be sufficiently long to insure flexibility to the system. The liquid level should never be allowed to rise to the pipes *M* and *N*, as any liquid other than that vertically over the drums will not be weighed correctly.

Tonnage Computed from Quantity of Refrigerant

The number of units of cooling effect available in the evaporation of one pound of ammonia under standard conditions has been found by equation [7] to be 465.5. The amount of ammonia required per ton is accordingly

$$\frac{200}{R\,b\,p - S(t_1 - t_2)} = \frac{200}{465.5} = 0.42964 \text{ pounds per minute.}$$

equivalent to 25.778 pounds per hour, or 618.7 pounds per twenty-four hours.

If, for example, it is found by test that 3,000 pounds of liquid ammonia per hour pass through a refrigerating system, the compression unit of which is a 19x38-inch double-acting compressor running at 45 revolutions per minute under standard conditions, the cooling effect produced is found to be

$$\frac{3000}{25.778} = 116.37 \text{ tons.}$$

Actual Displacement Efficiency of Compressor

Obviously, the efficiency of the compressor can be approximated by dividing the probable cooling effect, as determined by calculation based on the properties of the liquid, by the tonnage computed from the apparent displacement per minute in cubic feet as calculated from the dimensions of the compressor. Expressed as an equation this becomes

$$[11] \quad \text{Actual Efficiency} = \frac{\text{Actual Cooling effect}}{\text{Apparent Cooling effect}} = \frac{116.37}{144.66} = 80.4\%$$

Approximate Displacement Efficiency of Compressor

In the majority of cases the tonnage capacity of the system is required with reasonably close accuracy, but it is often impracticable to weigh the ammonia. In such cases a somewhat less accurate estimate of the efficiency of the compressor can be made with the assistance of an indicator. For all practical purposes the weight of ammonia gas may be considered proportional to its absolute pressure, and within narrow limits the amount of refrigeration represented by a cubic foot of ammonia gas will likewise be proportional to its absolute pressure. From this it follows that anything tending to reduce either the number of cubic feet of gas that a compressor handles or lower the pressure at which it is handled, proportionately reduces the capacity of the compressor. Graphically this is illustrated and the actual amount of the reduction in capacity is determined as follows:

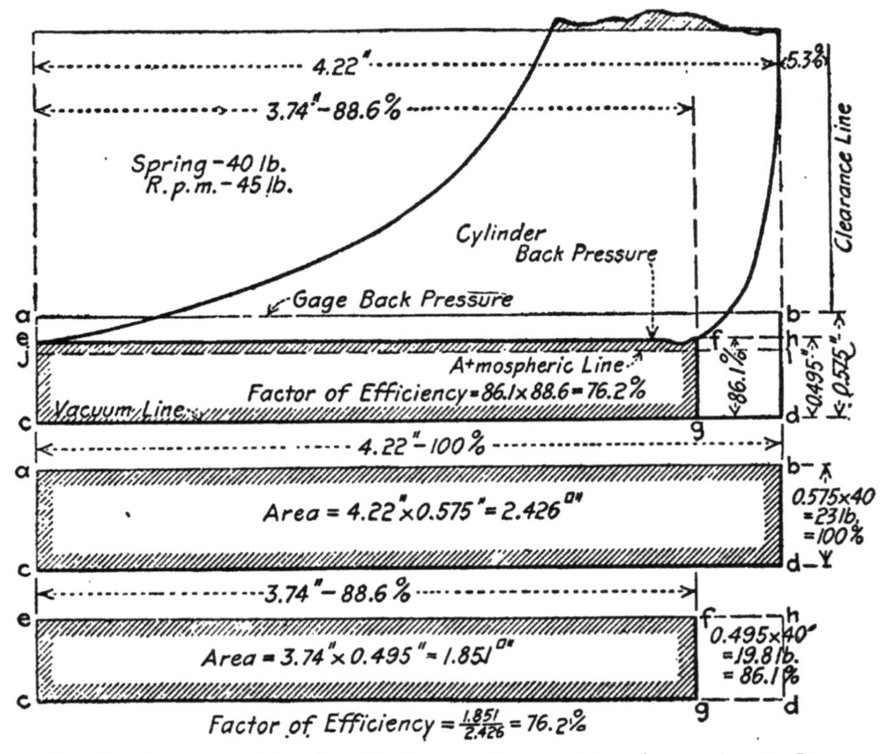

Fig. 42.—Diagram Showing Method of Determining Approximate Compressor Efficiency from Indicator Diagrams

Having taken an indicator diagram, such as that shown in Fig. 42, draw the lines *a b* and *c d* representing, respectively, the actual back pressure in the suction pipe, as indicated by a gauge, and the line of absolute vacuum. Next, determine *f*, the point at which the suction valve first opens to admit cold gas to the compressor cylinder. This point is the intersection of the admission line *e f* and the re-expansion line forming the heel of the diagram. Draw a vertical line *f g* through this point and other vertical lines *e c* and *b d* through the ends of the diagram. These horizontal and vertical lines form two rectangles. The larger one *a b d c* incloses a smaller one *e h d c* which, in turn, is made up of two still smaller rectangles *e f g c* and *f h d g*.

In the case under consideration the cylinder back-pressure line *a b* scales 4.8 pounds above the atmospheric line *j i*, making the absolute back pressure within the cylinder approximately 19.8 pounds. The observed suction pressure in the suction line is 8 pounds gage or approximately 23 pounds absolute, of which the 19.8 pounds is 86.1 per cent. This means that on account of the fall in back pressure, in passing through the suction valves and ports in entering the compressor cylinder, each cubic foot of gas represents only 86.1 per cent as much ammonia by weight as it would had no resistance been encountered and the cylinder back pressure been 23 pounds, the same as in the suction line.

The diagram shows that the compressor from which it was taken had excessive clearance. Due to the re-expansion of the high-pressure gases remaining in the clearance spaces, the opening of the suction valve is delayed until the piston has reached point *f* in the suction stroke. Cold returning ammonia gas can enter the compressor cylinder only during the time the piston is passing from *f* to the end of its stroke. The full length of the diagram represents the full stroke of the piston and a displacement of 100 per cent of the full volume of the cylinder. In this case, however, 11.4 per cent of the volume is occupied by re-expanding hot gas which reduces the amount of cold gas that can enter the compressor to 88.6 per cent. In other words, the actual displacement of the compressor in cubic feet is only 88.6 per cent of the apparent displacement, based on the cylinder dimensions only.

Now, the 88.6 per cent of the gas discharged weighs only 86.1 per cent as much as indicated by the pressure gauge on the suction line, so that the number of pounds of ammonia actually discharged by the compressor was only 88.6×86.1 per cent of what would be discharged by a compressor in which there is no clearance or resistance offered to the gas in passing through the suction valves.

Apparent Displacement

Graphically, the length *a b* of the large rectangle *a b d c* represents the compressor cylinder volume, and the height *b d* the absolute suction-gas pressure in the return line outside the compressor. The product of *a b* and *b d*, or cubic feet and weight per cubic foot,—since the weight of a gas depends upon the absolute pressure—represents the apparent displacement per stroke in pounds.

Actual Displacement

The length *e f* of the rectangle *e f g c* represents that part of the compressor-cylinder volume filled by the cold gas; and the high *g f*, the absolute suction pressure within the cylinder. The product of *e f* and *g f* represents the *actual* displacement per stroke in pounds, or the apparent displacement minus the re-expanded hot gas represented by the rectangle *f h d g*.

Approximate Displacement Efficiency

The approximate displacement efficiency of the compressor is represented by the ratio of the area of the small to the large rectangle and will be found to be numerically equal in this case to 76.2 per cent regardless of the units employed in measuring the areas. Using inches

$$\frac{3.74\times0.495}{4.22\times0.575}=\frac{1.851}{2.426}=76.2 \text{ per cent.}$$

or using pounds pressure and per cent stroke

$$\frac{19.8\times88.6}{23\times100}=76.2 \text{ per cent.}$$

Example:

The apparent number of cubic feet of ammonia gas discharged per minute by a 19x38-inch double-acting compressor running at 45 revolutions per minute has already been found to be 561.15. If it is found, by the indicator-diagram method just described, that the displacement efficiency is 76.2 per cent, the acutal number of cubic feet discharged will be

$$0.762\times561.15=427.59 \text{ cubic feet}$$

which, multiplied by the number of British thermal units of refrigeration represented in each cubic foot of ammonia gas actually displaced gives the total cooling effect of the machine expressed in British thermal units per minute. This quantity divided by 200 reduces the capacity to tons per 24 hours. Under standard conditions the number of British thermal units per cubic foot was found above to be 51.56, and, under standard conditions, the tonnage capacity of the compressor under consideration when operating at the determined efficiency of 76.2 per cent is

$$\frac{561.15\times 0.762\times 51.56}{200}=110.23 \text{ tons per 24 hours.}$$

Computation of Capacity

To expedite the figuring of capacities, not only under standard but also under other conditions, the accompanying tables of constants have been derived. To determine the tonnage capacity of a double-acting compressor of any size operating at any piston speed and under various conditions of head and back pressure it is necessary only to substitute appropriate values from these tables in the following equations.

Since 200 B.t.u. per minute is the equivalent of one ton per 24 hours, the tonnage capacity T of a compressor will be equal to the number of cubic feet of gas D actually displaced per minute multiplied by the number of British thermal units of cooling effect C available per cubic foot of gas actually displaced, divided by 200. But the actual displacement is equal to the apparent displacement D (figured from the dimensions and speed of the machine) multiplied by E, the displacement efficiency of the compressor, which may be assumed either from knowledge of the design, or calculated from the indicator diagrams in accordance with the above method, or determined by weighing the liquid refrigerant according to the method already described. Expressed as an equation, this becomes

$$T=\frac{C\,D\,E}{200} \qquad [12]$$

or since $\frac{200}{C}$ represents F the cubic feet required per ton

$$T=\frac{D\,E}{F} \qquad [13]$$

but

$$D=\frac{\frac{\pi}{4}\times d^2\times l\times r.p.m.\times 2}{1728}=d^2\times \text{piston speed}\times 0.00545418 \qquad [5]$$

from which the displacement can be computed readily when the diameter of the cylinder and the piston speed are known. The piston speed of a double-acting compressor of a given length of stroke, when operating at a given number of revolutions per minute, may also be determined directly from Table XII.

TABLE XII.—PISTON SPEED OF DOUBLE-

Length of Stroke in Inches	Revolutions per										
	1	24	26	28	30	31	32	33	34	35	36	37
1.....	0.1666	4	4.33	4.66	5	5.16	5.33	5.5	5.66	5.83	6	6.16
2.....	0.333	8	8.66	9.33	10	10.33	10.66	11	11.33	11.66	12	12.33
4.....	0.666	16	17.33	18.66	20	20.66	21.33	22	22.66	23.33	24	24.66
6.....	1	24	26	28	30	31	32	33	34	35	36	37
8.....	1.333	32	34.66	37.33	40	41.33	42.66	44	45.33	46.66	48	49.33
9.....	1.5	36	39	42	45	46.56	48	49.5	51	52.5	54	55.5
10.....	1.666	40	43.33	46.66	50	51.66	53.33	55	56.66	58.33	60	61.66
11.....	1.833	44	47.66	51.33	55	56.83	58.66	60.5	62.33	64.16	66	67.83
12.....	2	48	52	56	60	62	64	66	68	70	72	74
13.....	2.166	52	56.33	60.66	65	67.16	69.33	71.5	73.66	75.83	78	80.16
14.....	2.333	56	60.66	65.33	70	72.33	74.66	77	79.33	81.66	84	86.33
15.....	2.5	60	65	70	75	77.5	80	82.5	85	87.5	90	92.5
16.....	2.666	64	69.33	74.66	80	82.66	85.33	88	90.66	93.33	96	98.66
17.....	2.833	68	73.66	79.33	85	87.83	90.66	93.5	96.33	99.16	102	104.83
18.....	3	72	78	84	90	93	96	99	102	105	108	111
19.....	3.166	76	82.33	88.66	95	98.16	101.33	104.5	107.66	110.83	114	117.16
20.....	3.333	80	86.66	93.33	100	103.33	106.66	110	113.33	116.66	120	123.33
22.....	3.666	88	95.33	102.66	110	113.66	117.33	121	124.66	128.33	132	135.66
24.....	4	96	104	112	120	124	128	132	136	140	144	148
26.....	4.333	104	112.66	121.33	130	134.33	138.66	143	147.33	151.66	156	160.33
28.....	4.666	112	121.33	130.66	140	144.66	149.33	154	158.66	163.33	168	172.66
30.....	5	120	130	140	150	155	160	165	170	175	180	185
32.....	5.333	128	138.66	149.33	160	165.33	170.66	76	181.33	186.66	192	197.33
34.....	5.666	136	147.33	158.66	170	175.66	181.33	87	192.66	198.33	204	209.66
36.....	6	144	156	168	180	186	192	196	204	210	216	222
38.....	6.333	152	164.66	177.33	190	196.33	202.66	209	215.33	221.66	228	234.33
40.....	6.666	160	173.33	186.66	200	206.66	213.33	220	226.66	233.33	240	246.66
42.....	7	168	182	196	210	217	224	231	238	245	252	259
44.....	7.333	176	190.66	205.33	220	227.33	234.66	242	249.33	256.66	264	271.33
46.....	7.666	184	199.33	214.66	230	237.66	245.33	253	260.66	268.33	276	283.66
48.....	8	192	208	224	240	248	256	264	272	280	288	296
50.....	8.333	200	216.66	233.33	250	258.33	266.66	275	283.33	291.66	300	308.33
52.....	8.666	208	225.33	242.66	260	268.66	277.33	286	294.66	303.33	312	320.66
54.....	9	216	234	252	270	279	288	297	306	315	324	333
56.....	9.333	224	242.66	261.33	280	289.33	298.66	308	317.33	326.66	336	345.33
58.....	9.666	232	251.33	270.66	290	299.66	309.33	319	328.66	338.33	348	357.66
60.....	10	240	260	280	300	310	320	330	340	350	360	370
62.....	10.333	248	268.66	289.33	310	320.33	330.66	341	351.33	361.66	372	382.33
64.....	10.666	256	277.33	298.66	320	330.66	341.33	352	362.66	373.33	384	394.66
66.....	11	264	286	308	330	341	352	363	374	385	396	407
68.....	11.333	272	294.66	317.33	340	351.33	362.66	374	385.33	396.66	408	419.33
70.....	11.666	280	303.33	326.66	350	361.66	373.33	385	396.66	408.33	420	431.66
72.....	12	288	312	336	360	372	384	396	408	420	...	444

* From *Transactions* of the A. S. M. E.

It is also obvious that the apparent displacement is equal to the displacement per foot of piston travel multiplied by the number of feet of piston travel per minute. Table XIII gives this displacement per foot of piston travel, and Table XIV the number of cubic feet of ammonia that must be actually displaced per minute by the compressor to produce refrigeration at the rate of one ton per 24 hours.

Example: The tonnage capacity of a 19x38-inch double-

ACTING COMPRESSOR IN FEET PER MINUTE*

MINUTE

38	39	40	41	42	43	44	45	46	47	48	49	50
6.33	6.5	6.66	6.83	7	7.16	7.33	7.5	7.66	7.83	8	8.16	8.33
12.66	13	13.33	13.66	14	14.33	14.66	15	15.33	15.66	16	16.83	16.66
25.33	26	26.66	27.33	28	28.66	29.33	30	30.66	31.33	32	32.66	33.33
38	39	40	41	42	43	44	45	46	47	48	49	50
50.66	52	53.33	54.66	56	57.33	58.66	60	61.33	62.66	64	65.33	66.66
57	58.5	60	61.5	63	64.5	66	67.5	69	70.5	72	73.5	75
63.33	65	66.66	68.33	70	71.66	73.33	75	76.66	78.33	80	81.66	83.33
69.66	71.5	73.33	75.16	77	78.83	80.66	82.5	84.3	86.16	88	89.83	91.66
76	78	80	82	84	86	88	90	92	94	96	98	100
82.33	84.5	86.66	88.83	91	93.16	95.33	97.5	99.66	101.83	104	106.16	108.33
88.66	91	93.33	95.66	98	100.33	102.66	105	107.33	109.66	112	114.33	116.66
95	97.5	100	102.5	105	107.5	110	112.5	115	117.5	120	122.5	125
101.33	104	106.66	109.33	112	114.66	117.33	120	122.66	125.33	128	130.66	133.33
107.66	110.5	113.33	116.16	119	121.83	124.66	127.5	130.33	133.16	136	138.83	141.66
114	117	120	123	126	129	132	135	138	141	144	147	150
120.33	123.5	126.66	129.83	133	136.16	139.33	142.5	145.66	148.83	152	155.16	158.33
126.66	130	133.33	136.66	140	143.33	146.66	150	153.33	156.66	160	163.33	166.66
139.33	143	146.66	150.33	154	157.66	161.33	165	168.66	172.33	176	179.66	183.33
152	156	160	164	168	172	176	180	184	188	192	196	200
164.66	169	173.33	177.66	182	186.33	190.66	195	199.33	203.66	208	212.33	216.66
177.33	182	186.66	191.33	196	200.66	205.33	210	214.66	219.33	224	228.66	233.33
190	195	200	205	210	215	220	225	230	235	240	245	250
202.66	208	213.33	218.66	224	229.33	234.66	240	245.33	250.66	256	261.33	266.66
215.33	221	226.66	232.33	238	243.66	249.33	255	260.66	266.33	272	277.66	283.33
228	234	240	246	252	258	264	270	276	282	288	294	300
240.66	241	253.33	259.66	266	272.33	278.66	285	291.33	297.66	304	310.33	316.66
253.33	260	266.66	273.33	280	286.66	293.33	300	306.66	313.33	320	326.66	333.33
266	273	280	287	294	301	308	315	322	329	336	343	350
278.66	286	293.33	300.66	308	315.33	322.66	330	337.33	344.66	352	359.33	366.66
291.33	299	306.66	314.33	322	329.66	337.33	345	352.66	360.33	368	375.66	383.33
304	312	320	328	336	344	352	360	368	376	384	392	400
316.66	325	333.33	341.66	350	358.33	366.66	375	383.33	391.66	400	408.33	416.66
329.33	338	346.66	355.33	364	372.66	381.33	390	398.66	407.33	416	424.66	433.33
342	351	360	369	378	386	395	405	414	423	432	441	450
354.66	364	373.33	382.66	392	401.33	410.66	420	429.33	438.66	448	457.33	466.66
367.33	377	386.66	396.33	406	415.66	425.33	435	444.66	454.33	464	473.66	483.33
380	390	400	410	420	430	440	450	460	470	480	490	500
392.66	403	413.33	423.66	434	444.33	454.66	465	475.33	485.66	496	506.33	516.66
405.33	416	426.66	437.33	448	458.66	469.33	480	490.66	501.33	512	522.66	533.33
418	429	440	451	462	473	484	495	506	517	528	539	550
430.66	442	453.33	464.66	476	487.33	498.66	510	521.33	532.66	544	555.33	566.66
443.33	455	466.66	478.33	490	501.66	513.33	525	536.66	548.33	560	571.66	583.33
456	468	480	492	504	516	528	540	552	564	576	588	600

acting compressor of 76.2 per cent displacement efficiency operating at 45 revolutions per minute under 168 pounds head pressure and 16 pounds back pressure is

$$T = \frac{PSDE}{F} \qquad [14]$$

$$\frac{\left\{\begin{matrix}\text{Piston speed}\\ \text{from Table}\\ 12 = 285 \text{ feet}\\ \text{per minute}\end{matrix}\right\} \times \left\{\begin{matrix}\text{Apparent displacement}\\ \text{per foot of piston}\\ \text{travel from Table 13}\\ = 1.969 \text{ cubic feet}\end{matrix}\right\} \times \left\{\begin{matrix}\text{Displacement}\\ \text{efficiency of}\\ \text{compressor}\\ = 0.762\end{matrix}\right\}}{\text{Cu. ft. per min. per ton per 24 hours from Table XIV} = 3.85} = 111 \text{ tons}$$

TABLE XIII.—DISPLACEMENT *D* IN CUBIC FEET PER FOOT OF PISTON TRAVEL FOR VARIOUS-SIZED CYLINDERS

Cubic Feet per Inch of Piston Travel					Cubic Feet per Foot of Piston Travel				
Diameter, Inches and Fractions of Inch	0 Inch	¼ Inch	½ Inch	¾ Inch	Diameter, Inches and Fractions of Inch	0 Inch	¼ Inch	½ Inch	¾ Inch
1....	0.00045	0.00071	0.00102	0.00139	1....	0.00540	0.00852	0.01224	0.01668
2....	0.00182	0.00230	0.00284	0.00344	2....	0 02184	0.02766	0.03408	0.04128
3....	0.00409	0.00480	0.00557	0.00639	3....	0.04908	0.05760	0.06684	0.07668
4....	0.00727	0.00821	0.00920	0.01025	4....	0 08724	0.09852	0.11040	0.12300
5....	0.01136	0.01253	0.01375	0.01503	5....	0.13632	0.15036	0.16500	0.18036
6....	0.01636	0 01775	0.01920	0.02071	6....	0.19636	0.21300	0.23040	0.24852
7....	0 02227	0.02389	0.02557	0.02730	7....	0.26724	0.28668	0.30684	0.32760
8....	0.02909	0 03094	0.03284	0.03480	8....	0.34908	0.37128	0.39408	0.41760
9....	0.03682	0.03889	0.04102	0.04321	9....	0.44184	0.46668	0.49224	0.51852
10....	0.04545	0.04775	0.05011	0.05252	10....	0.54540	0.57300	0.60132	0.53024
11....	0 05500	0.05752	0.06011	0 06275	11....	0.66060	0.69024	0.72132	0.75300
12....	.06545	0.06821	0.07102	0.07389	12....	0.78540	0.81850	0.85224	0.88668
13....	0.07681	0.07980	0.08283	0.08593	13....	0.92172	0.95760	0.99396	1.03116
14....	.08908	0.09229	0 09556	0.09888	14....	1.0689	1.10748	1.14672	1.18556
15....	.10226	0.10570	0.10920	0.11275	15....	1.2271	1.26840	1.31040	1.35300
16....	0.11636	.12002	0.12374	0.12752	16....	1.3963	1.44024	1.48488	1.53021
17....	0.13135	.13525	0.13919	0.14320	17....	1.5762	1.62300	1.67028	1.71840
18....	0.14726	0.15138	0.15556	0 15979	18....	1.7671	1.81650	1.86672	1.91748
19....	0.16408	.16843	0.17283	0.17729	19....	1.9689	2.02116	2.07396	2.12748
20....	0.18181	.18638	0.19101	0.19570	20....	2.1817	2.13656	2.29212	2.34840
21....	0 20044	20524	0.21010	0.21501	21....	2.4053	2.46288	2.52120	2.58012
22....	0.21998	.22501	2 23010	0.23524	22....	2.6397	2.70072	2.76120	2.82280
23....	0.24044	0.24569	0.25100	0.25637	23....	2.8852	2.94828	3.01200	3.07644
24....	0.26180	26728	0.27282	0.27842	24....	3.1416	3.20736	3.27364	3.34104
25....	0.28407	.28978	0.29555	0.30137	25....	3.4088	3.47736	3.54666	3.61644
26....	0.30725	0 31319	0 31918	0.32523	26....	3.6870	3.75828	3.83016	3.90276
27....	0.33134	0 33750	0 34373	0.35000	27....	3.9760	4.05000	4.12476	4.20000
28....	0.35634	0 36273	0.36918	0.37568	28....	4.2760	4.35276	4.43016	4.50816
29....	0.38225	0.38886	0.39554	0.40227	29....	4.5870	4.66632	4.74648	4.82524
30....	0.40906	0.41591	0 42281	0.42977	30....	4.9081	4.99092	5.07372	5.15724

Cooling Effect Produced on Brine

A method which avoids opportunity for error in determining the compressor displacement efficiency, but at the same time introduces another difficulty in the form of Brine-tank insulation losses—which fortunately can usually be more or less accurately determined and corrected for—is to check the apparent performance of the compressor by the actual performance of the refrigerating system as a whole, as determined by direct measurement of the cooling effect produced on brine, where it is regularly employed in the plant, or on brine, water, or some other fluid of known specific heat where a secondary medium has to be introduced for the purpose of test. If, in the latter case, the refrigerating effect cannot be put to useful work, it may be neutralized by artificial

TABLE XIV.—CUBIC FEET F AND POUNDS P OF AMMONIA PER TON OF REFRIGERATION PER 24 HOURS

W = Weight per cubic foot BP = Back pressure			Head Pressure, Condenser or Guage Pressure and Corresponding Temperature											
			100 Pounds. 63.5 Degrees	110 Pounds. 68 Degrees	120 Pounds. 72.6 Degrees	130 Pounds. 77.4 Degrees	140 Pounds. 80.3 Degrees	150 Pounds. 83.8 Degrees	160 Pounds. 87.4 Degrees	170 Pounds. 90.8 Degrees	180 Pounds. 93.8 Degrees	190 Pounds 96.9 Degrees	200 Pounds. 100 Degrees	Temperature, Degrees Fahrenheit
W BP	0.0556 0	P F	0.4159 7.482	0.4199 7.551	0.4240 7.626	0.4284 7.703	0.4310 7.761	0.4343 7.812	0.4376 7.870	0.4408 7.929	0.4440 7.986	0.4470 8.041	0.4501 8.095	−28.5
W BP	0.0133 5	P F	0.4122 5.636	0.4160 5.675	0.4202 5.732	0.4243 5.790	0.4271 5.826	0.4308 5.878	0.4335 5.914	0.4366 5.970	0.4397 5.999	0.4437 6.039	0.4458 6.081	−17.5
W BP	0.0910 10	P F	0.4093 4.502	0.4130 4.543	0.4171 4.587	0.4204 4.625	0.4237 4.662	0.4271 4.698	0.4302 4.733	0.4332 4.766	0.4363 4.799	0.4392 4.833	0.4423 4.865	− 8.5
W BP	0.1083 15	P F	0.4068 3.756	0.4106 3.791	0.4145 3.827	0.4186 3.866	0.4211 3.889	0.4244 3.918	0.4276 3.948	0.4288 3.975	0.4336 4.003	0.4365 4.030	0.4394 4.058	− 1
W BP	0.1258 20	P F	0.4040 3.211	0.4077 3.241	0.4116 3.272	0.4158 3.305	0.4182 3.324	0.4214 3.350	0.4245 3.375	0.4275 3.398	0.4304 3.422	0.4333 3.444	0.4362 3.467	5.66
W BP	0.1429 25	P F	0.4025 2.819	0.4062 2.843	0.4102 2.870	0.4140 2.898	0.4167 2.916	0.4198 2.938	0.4229 2.959	0.4258 2.980	0.4287 3.000	0.4316 3.020	0.4345 3.040	11.5
W BP	0.1600 30	P F	0.4013 2.507	0.4049 2.530	0.4088 2.555	0.4128 2.580	0.4152 2.600	0.4184 2.615	0.4213 2.633	0.4243 2.653	0.4273 2.671	0.4300 2.687	0.4329 2.706	16.8
W BP	0.1766 35	P F	0.3991 2.260	0.4028 2.280	0.4066 2.302	0.4105 2.925	0.4130 2.338	0.4161 2.356	0.4188 2.373	0.4220 2.390	0.4249 2.406	0.4277 2.422	0.4305 2.443	21.7
W BP	0.1941 40	P F	0.3984 2.052	0.4020 2.071	0.4058 2.090	0.4098 2.111	0.4122 2.123	0.4153 2.139	0.4183 2.155	0.4211 2.175	0.4240 2.185	0.4269 2.200	0.4296 2.214	26.1

$$F = \frac{144 \times 2000}{W[1440\, R_{B.P.} - S(t_1 - t_2)]} \qquad P = \frac{144 \times 2000}{1440\, R_{B.P.} - S(t_1 - t_2)}$$

heat introduced through the agency of a steam coil or electrical resistance.

The amount of cooling that a given quantity of brine will do depends not only upon the number of degrees rise in temperature, but upon the density and kind of brine.

The most important element in the selection of the kind of brine to use is the temperature to be produced, which fixes the temperatures at which the brine must be circulated. Saturated salt brine, by which is meant brine so strong that it will dissolve no more salt, freezes at about 5° Fahrenheit below zero and would be safe for brine-tank temperatures above zero.

The weaker the brine the higher the temperature at which it freezes, the limit being reached when the amount of salt is reduced to nothing, in which case the brine becomes water and freezes at 32° Fahrenheit.

Saturated calcium brine freezes at about 55° Fahrenheit below zero and according to its densities is adapted to brine temperatures from 40° below zero up. The specific heat of either salt or calcium brine upon which depend their refrigerating capacities per pound per degree rise in temperature, decreases as the strength increases.

The refrigerating capacity of water per pound per degree rise in temperature is one British thermal unit. As salt or calcium chloride is added to the water this value decreases until its value at saturation (maximum strength) is only 0.77 B.t.u. In the latter case, about 30 per cent more brine must be circulated, to accomplish a given amount of cooling for a given rise in brine temperature, than would be necessary were the desired temperatures sufficiently high to allow water to be employed instead of brine as the medium for conveying heat.

Approximate Cooling Effect Twenty-five Heat Gallons per Ton

For ordinary accuracy 25 "heat gallons" is considered equivalent to a ton of cooling effect, a "heat gallon" being the cooling effect required to reduce the temperature of 1 gallon of calcium chloride brine, of 1.2 specific gravity, through a range of 1° Fahrenheit per minute.

The volume of brine circulated expressed in cubic feet per minute can be calculated from the dimensions and strokes per minute of the pump, due allowance being made for slippage, and this can be readily converted into weight by multiplying by the weight of brine per cubic foot as given in the accompanying tables.*

TABLE XV.—PROPERTIES OF *CALCIUM* CHLORIDE BRINE

Salt Required		Freezing Point		Amm. Gauge Pressure Lbs.	Degrees Salinometer at 60°F.	Degrees Baumé at 60°F.	Specific Gravity at 60°F.	Specific Heat	Percentage of Salt by Weight
Lbs. per Gallon	Lbs. per Cu. Ft.	Degrees F.	Degrees C.						
½	3¾	+29	−1.6	43	12	3	1.024	0.980	3
1	7½	+27	−2.8	39	27	6	1.041	0.964	5
1¼	8½	+25	−3.9	37	36	9	1.058	0.936	7
1½	11¼	+23	−5.0	35½	40	10	1.076	0.911	9
1¾	13	+21	−6.1	34	44	11	1.085	0.896	10
2	15	+18	−7.8	30½	52	13	1.103	0.884	12
2¼	17	+14	−10.0	26	62	15	1.121	0.868	14
2½	19	+4	−15.5	18	80	20	1.159	0.844	18
3	22½	−1.5	−18.2	12½	88	22	1.179	0.834	20
3½	26	−8	−22.2	8	95	24	1.199	0.817	22
4	30	−17	−27.2	4	104	26	1.219	0.799	24
4½	34	−27	−32.8	1" Vacuum	112	28	1.240	0.778	26
5	37½	−39	−39.4	8" "	120	34	1.305	.....	32
5½	41	−54	−47.7	15" "	Max. D	en., 32	1.283	.....	30

TABLE XVI.—PROPERTIES OF *SODIUM* CHLORIDE BRINE

Salt Required		Freezing Point		Amm. Gauge Pressure Lbs.	Degrees Salinometer at 60°F.	Degrees Baumé at 60°F.	Specific Gravity at 60°F.	Specific Heat	Percentage of Salt by Weight
Lbs. per Gallon	Lbs. per Cu. Ft.	Degrees F.	Degrees C.						
0.084	0.63	+30.5	−0.8	45	4	1	1.007	0.992	1
0.169	1.26	+29.3	−1.5	43.5	8	2	1.015	0.984	2
0.212	1.58	+28.6	−1.9	42.5	10	3	1.019	0.980	2.5
0.256	1.92	+27.8	−2.3	42	12	3.5	1.023	0.976	3
0.300	2.24	+27.1	−2.7	41.5	14	4	1.026	0.972	3.5
0.344	2.57	+26.6	−3.0	40	16	4.5	1.030	0.968	4
0.433	3.24	+25.2	−3.8	39	20	5.5	1.037	0.960	5
0.523	3.92	+23.9	−4.5	37.6	24	6.5	1.045	0.946	6
0.617	4.63	+22.5	−4.7	35.5	28	7.6	1.053	0.932	7
0.708	5.3	+21.2	−5.3	34.5	32	8.7	1.061	0.919	8
0.802	6.0	+19.9	−6.7	33	36	9.7	1.068	0.905	9
0.897	6.7	+18.7	−7.4	31.5	40	10.7	1.076	0.892	10
1.092	8.2	+16.0	−8.9	29.2	48	12.6	1.091	0.874	12
1.389	10.4	+12.2	−11.0	25.5	60	15.7	1.115	0.855	15
1.928	14.4	+6.1	−14.7	20.3	80	20.4	1.155	0.829	20
2.376	17.78	+1.2	−16.5	16.5	96	24	1.187	0.795	24
2.488	18.68	+0.5	−17.3	16	100	25	1.196	0.783	25
2.610	19.5	−1.1	−18.2	14.8	...	25.8	1.204	0.771	26
.....	.....	−4.7	−20.1	12.5	...	....	.....	.....	29

Example:—It is found by test that the brine pump of a brine-circulating system discharges 1,000 gallons of brine per minute, the temperature of the warm return brine being 7° Fahrenheit

* The specific gravity of a substance is the ratio of the weight of that substance to the weight of the same volume of pure water at its temperature of maximum density at 39 degrees Fahrenheit, at which temperature it weighs 62.425 pounds per cubic foot. The weight per cubic foot of brine given in the table is determined by multiplying 62.425 by the specific gravity as determined by a salinometer or other similar hydrometric instrument.

higher than the outgoing cold brine. One thousand gallons and 7° rise is equivalent to 7,000 heat gallons, which, divided by 25, the approximate number of heat gallons per ton, gives 280 as the approximate tonnage capacity at which the system is operating.

This rule is intended to apply roughly to brines of the higher densities, and, since it does not take into consideration possible variations in the value of the specific heat of the brine, it cannot be expected to apply accurately to brines of all densities. For example, according to formula [15] the amount of refrigeration produced by the circulation of 200 pounds of water per minute with a rise in temperature of 1° Fahrenheit would be

$$T = \frac{200 \times 1 \times 1}{200} = 1 \text{ ton.}$$

According to the rule, which ignores the specific heat of water, which is unity, the cooling effect would be, since 200 pounds of water is 24 gallons,

$$\frac{\text{(Gals. per min.)}}{\text{(No. of heat gals. per ton)}} = \frac{24}{25} = 0.96 \text{ ton}$$

Actual Cooling Effect

If we have the specific information that the brine is of a density of 120° salinometer, corresponding to a specific gravity of 1.305, weighing $1.305 \times 62.425 = 81.464$ pounds per cubic foot, and that the specific heat of the brine of this density is .767, the cooling effect expressed in tons per 24 hours will be found by the following equation:

$$[15] \quad T = \frac{W\,S\,(t_1 - t_2)}{200} \text{ in which}$$

W = Weight of brine circulated per minute.

S = Specific heat of the brine.

$(t_1 - t_2)$ = Range in temperature cooled through.

200 = Number of B.t.u. per minute equivalent to a ton of refrigeration per 24 hours.

Since there are 7.48 United States gallons in a cubic foot, the weight of the brine per gallon is

$$1.305 \times 62.425 \times 7.48 = 10.89$$

pounds, and since 1,000 gallons is circulated per minute the weight (W) to be substituted in the foregoing expression is 10,890. From the table we find that the specific heat of brine of 1.305 specific gravity is 0.767 and since the range through which the brine is cooled is 7° Fahrenheit, we have

$$Tons = \frac{10.890 \times 0.767 \times 7}{200} = 292.34 \text{ tons.}$$

For accurate determinations of the cooling effect the density of the brine should be determined by either a salinometer or some other form of hydrometer that will allow either the percentage of saturation or the specific gravity of the brine to be determined. In taking such hydrometer readings care should be taken to bring the temperature of the brine to that at which the instrument is calibrated. This method is less likely to lead to error than that of applying a correction factor for reducing the readings taken at other temperatures to what they would be if taken at the standard temperature.

For very accurate determinations the amount of brine cooled should be determined by weighing and its specific heat determined by some competent expert. Thermometers used in taking the temperatures of the secondary medium before and after cooling, as well as all other apparatus, should be carefully calibrated, both before and after the test.

CHAPTER XI

COLD STORAGE DUTY

It is obviously impossible to determine with any great degree of accuracy how much refrigeration it will take to cool a number of differently shaped cold storage boxes, built by unknown methods of construction, of several different insulating materials of unknown efficiencies and various states of preservation, into which heat is admitted through the opening of doors and radiated from lights and workmen, and containing unknown quantities of different kinds of products of varying heat absorbing capacities, stored for different lengths of time.

When the above list of unknown quantities can be sufficiently reduced, however, calculations of the amount of cold storage capacity required to satisfy a certain set of conditions can readily be made. Since lights and workmen must necessarily be employed to a greater or less extent, and since no insulation can entirely prevent the inflow of heat, only a part of the refrigeration produced by the refrigerating plant can be put to the useful work of cooling the stored products, the remainder being dissipated.

Attempts are sometimes made to estimate cold storage duty by determining the number of cubic feet of space to be cooled and dividing that by the number of cubic feet that a ton of refrigeration is supposed to cool under average conditions. While it may be interesting to know the amount of space cooled by a ton of refrigeration under more or less similar conditions, such comparisons are not only meaningless but are positively misleading, when the many varying conditions of operation are not definitely known. Since such calculations are at best inaccurate, they should be made only on the basis of the greatest number of known quantities and carefully worked out assumptions regarding the remaining unknown quantities.

These determinations may be simplified by the following brief method, which, together with the accompanying series of tables, will, when judiciously applied, be found accurate enough for all ordinary commercial requirements.

The total amount of heat that the refrigerating machine must remove from the cold storage compartment, is made up as follows:

a. Latent and sometimes specific heat of the products stored.
b. Heat evolved by lights.
c. Heat given off by workmen.
d. Heat absorbed in the precipitation and freezing of moisture.
e. Heat of air entering through open doors.
f. Heat entering through the cold storage insulation.

Cooling the Product

The amount of refrigeration required to cool a given amount of food product through a given range in temperature is a practically fixed quantity for a given product, but varies widely with different products. When cooling is not to be carried below the freezing point the amount of refrigeration required may be found by multiplying the specific heat of the product by the number of degrees through which it is to be cooled. If the product is also to be frozen, this amount of refrigeration must be increased by the amount of the latent heat of fusion, and if cooling is to be continued below the freezing point, the refrigeration must be further increased by the specific heat of the product below 32° multiplied by the number of degrees through which it is cooled below freezing point. The specific and latent heat of a number of products commonly preserved in cold storage are given in the following table:

TABLE XVII—PROPERTIES OF FOOD PRODUCTS

Product	Specific Heat Above 32°F.	Latent Heat of Freezing	Specific Heat Below 32°F.
Beef—lean	.77	102	.41
Beef—fat	.60	72	.34
Butter	.64	...	.84
Cream	.68	84	.38
Eggs	.76	100	.40
Fish	.82	111	.43
Milk	.90	124	.47
Mutton	.67	...	.84
Oysters	.84	114	.44
Poultry	.80	105	.42
Pork—fat	.51	55	.30

Example.—It is required to cool 10,000 pounds of freshly killed poultry through 68° Fahr. The specific heat as given above is 0.80. The number of B. t. u. to be removed will be $0.80 \times 10{,}000 \times 68 = 544{,}000$. Dividing this result by 144 (number of B. t. u. per pound of refrigeration), the amount of cooling duty is found to be 3,777.7 pounds. If the poultry is frozen, the additional refrigeration required will be 10,000×105=1,050,000 B. t. u. or (÷144) 7,292 pounds, and if additional cooling to 0° Fahr. is required, the additional cold necessary will be 10,000×0.42×32=134,000 B. t. u. or 933.3 pounds. The total refrigeration duty required to cool the products through 68° Fahr., freeze it at 32° Fahr., and then chill it to 0° Fahr., would be 3,777.7+7,292+933.3=12,003 pounds or dividing by 2,000 (pounds per ton), 6 tons.

The following table may be found convenient in estimating the amount of refrigeration required to chill beef, pork, and sausage through 64° Fahr., or from 104° Fahr. to 40° Fahr.:

TABLE XVIII—REFRIGERATION REQUIRED TO COOL MEATS

Products—	Poultry	Beef Fat	Beef Medium	Beef Lean	Pork Fat	Sausage (15% Water)
Specific Heat	.80	.60	.68	.77	.51	.65
B.t.u. to cool 1,000 pounds 1°Fahr	800	600	680	770	510	650
B.t.u. to cool 1,000 pounds 64°Fahr	51200	38400	43520	49280	32640	41600
Pounds Refrigeration per 1,000 pounds (64°Fahr.)	355.55	266.66	302.22	333.66	226.66	228.88
Pounds of Meat cooled 64° per ton Refrigeration	5,625	7,500	6,615	5,844	8,765	6,923
Average wt. Carcass	....	750 lbs.	750 lbs.	750	250	....
Carcasses cooled per ton	....	10	8.82	7.78	35.3	....

It will be noted that ten 750 pound fat beeves, and thirty-five 250 pound hogs require one ton of refrigeration for the cooling of the meat alone. In estimating the cooling capacity of a medium for packing house work, a ton of refrigeration is allowed for from five to seven beeves weighing from 700 to 750 pounds, and for from fifteen to twenty-four hogs weighing 250 pounds. Still another rough rule sometimes employed is to allow a ton of refrigeration for from 3 to 4,000 pounds of meats cooled. These larger figures are intended to give ample reserve capacity to provide for ordinary insulation and other losses encountered in packing house practice.

Water Cooling

Since the specific heat of water is unity, the number of heat units to be extracted in order to produce a given drop in temperature of a given quantity of water is found by simply multiplying the weight in pounds by the range cooled through in degrees.

If, for example, 20,000 pounds of water is to be cooled one degree, 1,000 pounds 20 degrees, 400 pounds 50 degrees, or in fact any number of pounds through a range of temperature that will give a product 20,000 pound degrees, 20,000 B. t. u. will be required for the cooling.

One U. S. gallon of water at 62° F. weighs 8.336 pounds. The cooling of 20,000 gallon degrees will accordingly require $8.336 \times 20,000$ or 166,720 B. t. u. If the cooling is accomplished in 24 hours the amount of refrigeration required will be $166,720 \div 288,000 = .5789$ tons. If done in one hour the equivalent rate per 24 hours will be 24 times as great or $.5789 \times 24 = 13.893$ tons.

Table XIX shows the amount of refrigeration required to cool 1,000 gallons of water per minute, hour and 24 hours through different ranges of temperature. If 1,000 gallons of water be cooled 50 degrees in one hour, the equivalent cooling effect per 24 hours will be ten times the value given in the table for 5 degrees, or 34.733 tons; if 53 degrees, ten times the value for 5 degrees, or 34.733 plus that given for 3 degrees or 2.0840, making 36.817 tons.

TABLE XIX—REFRIGERATION REQUIRED TO COOL WATER — GALS. — DUTY IN TONS PER 24 HOURS

Degrees Cooled	1,000 Gals. Cooled per			Degrees Cooled	1,000 Gals. Cooled per		
	Minute	Hour	24 Hours		Minute	Hour	24 Hours
1	41.68	0.6946	.02894	21	875.28	14.5879	.60778
2	83.36	1.3893	.05789	22	916.96	15.2824	.61672
3	125.04	2.0840	.08682	23	958.64	15.9770	.65564
4	166.72	2.7786	.11577	24	1000.32	16.6716	.69456
5	208.40	3.4733	.14471	25	1042.00	17.3664	.72352
6	250.08	4.1679	.17364	26	1083.68	18.0612	.75248
7	291.76	4.8646	.20259	27	1125.36	18.7598	.78142
8	333.44	5.5590	.23154	28	1167.04	19.4584	.81036
9	375.12	6.2519	.26048	29	1208.72	20.1491	.83931
10	416.80	6.9466	.28942	30	1250.40	20.8399	.86826
11	458.48	7.6412	.30836	31	1292.08	21.5379	.89721
12	500.16	8.3358	.34728	32	1333.76	22.2360	.92616
13	541.84	9.0306	.37624	33	1375.44	22.9289	.95510
14	583.52	9.7292	.40518	34	1417.12	23.6218	.98404
15	625.20	10.4199	.43413	35	1458.80	24.3147	1.01298
16	666.88	11.1180	.46308	36	1500.48	25.0076	1.04192
17	708.56	11.8109	.49202	37	1542.16	25.7023	1.07086
18	750.24	12.5038	.52096	38	1583.84	26.3970	1.09980
19	791.92	13.1985	.54990	39	1625.52	27.0918	1.12874
20	833.60	13.8933	.57884	40	1667.20	27.7866	1.15768

WORT COOLING

Table XX shows the amount of refrigeration expressed in tons per 24 hours required to cool 100 bbls. of water per hour and one bbl. per minute through different ranges of temperature. To apply this table to wort cooling multiply the number of bbls. of 31 gallons by the specific gravity of the wort and this product by the specific heat of the wort corresponding to the specific gravity as shown in Fig. 43.

TABLE XX—REFRIGERATION REQUIRED TO COOL WATER (BBLS. OF 31 GALS.) DUTY IN TONS OF REFRIGERATION PER 24 HOURS

Degrees Cooled	100 Bbls. of 31 Gals., per Hour	1 Bbl. of 31 Gals., per Min.	Degrees Cooled	100 Bbls. of 31 Gals., per Hour	1 Bbl. of 31 Gals., per Min.
1	2.1545	1.2927	21	45.2445	27.1467
2	4.3090	2.5854	22	47.3990	28.4394
3	6.4635	3.8781	23	49.5535	29.7331
4	8.6180	5.1708	24	51.6988	31.0248
5	10.7725	6.1635	25	53.8625	32.3175
6	12.9270	7.7562	26	56.1370	33.6102
7	15.0815	9.0489	27	58.1715	34.9029
8	17.2360	10.3416	28	60.3260	36.1956
9	19.3905	11.6343	29	62.4805	37.4883
10	21.5950	12.9270	30	64.6350	38.7810
11	23.6995	14.2197	31	66.7895	39.0737
12	25.8494	15.5124	32	68.9440	41.3664
13	28.0685	16.8051	33	71.0985	42.6591
14	30.1630	18.0978	34	73.0530	43.9518
15	32.3175	19.3905	35	75.4075	45.2445
16	34.4720	20.6832	36	77.5620	46.5372
17	36.5265	21.9759	37	79.7165	47.8299
18	38.7810	23.2686	38	81.8710	44.1226
19	40.9355	24.5613	39	83.9255	50.4153
20	43.0900	25.8540	40	86.1800	51.6581

TABLE XXI—PRODUCTS OF SPECIFIC GRAVITY AND SPECIFIC HEAT OF WORT OF DIFFERENT PER CENT STRENGTH

Strength %	Product	Strength	Product
8	0.9742	15	0.9499
9	0.9741	16	0.9463
10	0.9665	17	0.9426
11	0.9631	18	0.9390
12	0.9609	19	0.9353
13	0.9571	20	0.9320
14	0.9536	..	......

Refrigeration required to cool wort = that required to cool equal quantity of water, multiplied by the above "product" corresponding to strength of wort.

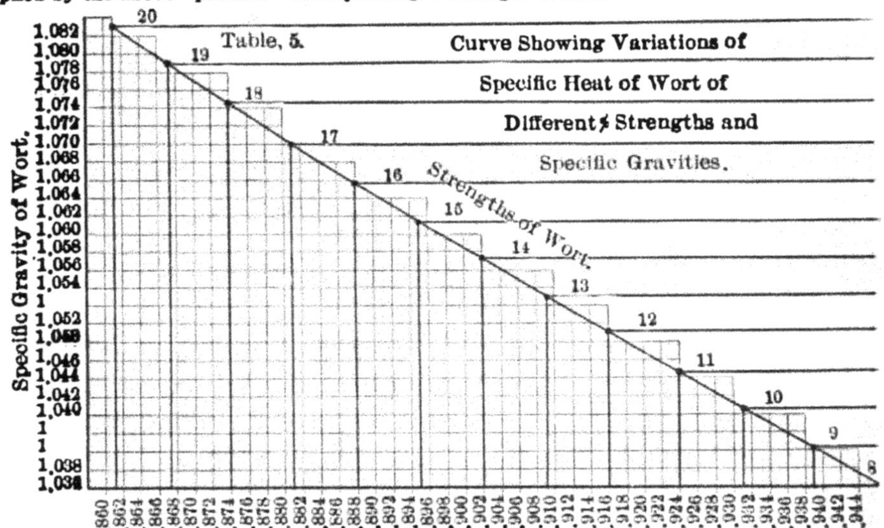

Fig. 43

Suppose it is desired to find the amount of refrigeration necessary to cool 100 bbls. of wort per hour having a strength of 12% through a range of 40° F. It is found from table 20, the refrigeration required to cool a like amount of water is 86.18 tons. Since the specific gravity of the wort, as determined from Fig. 43, is 1.049, the weight of the wort cooled will be 1.049 times as great as for water; but since the specific heat is only .916 the refrigeration per pound will be only .916 as great. The product of these two factors .9609—given for different strengths of wort in Table XXI—shows that the amount of refrigeration required to cool a given quantity of wort of 12% strength is .9609 as great as for the same quantity of water or in the above example of 100 bbls. per hour, .9609×86.18 or 82.80 tons.

The same result might have been obtained from Table XIX by first reducing the quantity in bbls. to gallons, or direct from the following equation in which for clearness the above values have been substituted.

$$\frac{\begin{pmatrix}\text{Bbls. of}\\ \text{wort per}\\ \text{hour}\\ B=100\end{pmatrix}\begin{pmatrix}\text{Gals.}\\ \text{per bbl.}\\ \text{wort}\\ G=31\end{pmatrix}\begin{pmatrix}\text{Wt. water}\\ \text{per gal.,}\\ \text{lbs.}\\ L=8.34\end{pmatrix}\begin{pmatrix}\text{No. de-}\\ \text{grees F.}\\ \text{cooled}\\ (t_1-t_2)=40\end{pmatrix}\begin{pmatrix}\text{Specific}\\ \text{gravity}\\ \text{of wort}\\ Sq=1.049\end{pmatrix}\begin{pmatrix}\text{Specific}\\ \text{heat of}\\ \text{wort}\\ Sp=.916\end{pmatrix}}{(\text{B.t.u. per hour equivalent to a ton of refrigeration per 24 hours}=12000)}=\begin{pmatrix}\text{Rate of}\\ \text{cooling.}\\ \text{Tons per}\\ \text{24 hrs.}\\ T=82.80\end{pmatrix} \quad [16]$$

This expression when applied to wort cooling expressed in bbls. of 31 gallons cooled per hour, becomes:

$\text{Tons} = .021545 \times (t - t_1) \times sg. \times sh.$ in which

$(t - t_1)$ = range of temperature cooled through. (40° F.)

$Sg.$ = specific gravity of the wort. (1.049)

$Sh.$ = specific heat of the wort. (.916)

which values substituted in the above equation give tons per 24 hours = 82.80 as above.

Lights

The heat generated by artificial lights in cold storage compartments often becomes of considerable importance. The amount of refrigeration required to neutralize the heat radiated by electric lights depends on the candle power of the lights and their efficiency, expressed in watts per candle power. If, for example, an ordinary low efficiency lamp consuming 3.5 watts per candle power is employed, the heat per 16 c. p. will be 16×3.5×0.05685=3.1836 B.t.u. per minute or 191.0 B.t.u. per hour, which quantities divided respectively by 0.1 and 6.0, the number of B.t.u. per minute and hour equivalent to a pound of refrigerating duty per twenty-four hours, gives 31.83 pounds. One ton of refrigeration equivalent to 2,000 pounds of ice melting capacity will accordingly be required for every sixty-three lights.

The heat generated by an ordinary gas light is 4,800 B. t. u. per hour, to absorb which requires 800 pounds of refrigeration per twenty-four hours ($24 \times 4800 \div 144 = 800$). On this basis two and one-half gas lights will absorb a ton of refrigeration per twenty-four hours ($2000 \div 800 = 2\frac{1}{2}$).

Similarly each workman employed in cold storage compartments radiates about 500 B. t. u. per hour, equivalent to $83\frac{1}{3}$ pounds of refrigeration ($24 \times 500 \div 144 = 83\frac{1}{3}$). On which basis a ton of refrigeration will have to be supplied for every 24 workmen per twenty-four hours ($2000 \div 83\frac{1}{3} = 24$). In commercial practice some authorities allow a ton of refrigeration for every 30 workmen, presupposing a radiation of 400 B. t. u. per man.

Air Cooling and Moisture Precipitation

Before proceeding to illustrate the method of calculating the amount of refrigeration required to cool a mixture of air and water vapor it may be advisable to define terms.

Air is a mechanical mixture of nitrogen and oxygen in the practically constant proportion of 80 parts of the former to 20 parts of the latter, a very small per cent., about 3 or 4 hundredths of one per cent. of which is replaced by carbon dioxide. Into this uniform mechanical mixture water vapor enters in widely varying proportions. When the air contains all the moisture that it can hold it is said to be saturated. The higher the temperature of the air the more water vapor it is capable of absorbing before becoming saturated. At a given temperature saturated air always contains a certain fixed quantity of water vapor.

It must be remembered, however, that the temperature of the air does not fix the amount of moisture that it contains except in the limiting case of saturation. In the general case the air is not saturated, and may contain different amounts of water at the same temperature as it varies in degree of saturation; or it may contain different amounts of moisture at the same degree of saturation at different temperatures. Since the amount of water that it is possible for air to hold in suspension increases with increasing temperature, and decreases with decreasing temperature, it is evident that the air may or may not contain less moisture after

TABLE XXII—TABLE SHOWING POUNDS OF AQUEOUS VAPOR IN 1,000 CU. FT. OF AIR AT DIFFERENT TEMPERATURES AND PER CENT. SATURATION

At	10%	15%	20%	25%	30%	35%	40%	45%	50%	55%	60%	65%	70%	75%	80%	85%	90%	95%	100%	At
100°F.	.282	.423	.556	.705	.847	.988	1.130	1.270	1.411	1.552	1.694	1.835	1.976	2.117	2.259	2.400	2.541	2.682	2.823	100°F.
95°F.	.244	.366	.489	.621	.733	.855	.978	1.100	1.223	1.345	1.467	1.589	1.712	1.844	1.977	2.089	2.201	2.323	2.446	95°F.
90°F.	.211	.316	.422	.527	.633	.740	.847	.951	1.056	1.161	1.267	1.373	1.479	1.584	1.690	1.795	1.901	2.005	2.110	90°F.
85°F.	.182	.272	.363	.454	.545	.636	.727	.818	.909	1.000	1.091	1.182	1.273	1.364	1.455	1.546	1.637	1.728	1.819	85°F.
80°F.	.156	.234	.312	.390	.468	.546	.624	.702	.781	.859	.937	1.015	1.093	1.166	1.249	1.327	1.405	1.483	1.562	80°F.
75°F.	.133	.200	.267	.334	.401	.467	.534	.601	.668	.735	.802	.868	.935	1.002	1.069	1.135	1.202	1.269	1.336	75°F.
70°F.	.114	.171	.228	.285	.342	.399	.456	.513	.570	.627	.684	.741	.798	.855	.912	.964	1.026	1.083	1.140	70°F.
65°F.	.096	.144	.193	.241	.290	.338	.387	.435	.484	.532	.581	.629	.678	.726	.775	.823	.872	.920	.968	65°F.
60°F.	.082	.123	.164	.205	.246	.286	.327	.368	.410	.451	.492	.533	.574	.615	.656	.697	.738	.779	.820	60°F.
55°F.	.069	.103	.138	.172	.207	.242	.277	.311	.346	.380	.415	.449	.484	.519	.554	.588	.623	.657	.692	55°F.
50°F.	.058	.087	.116	.145	.174	.204	.232	.261	.291	.320	.349	.378	.407	.436	.465	.494	.524	.553	.582	50°F.
45°F.	.048	.072	.097	.121	.146	.170	.195	.219	.243	.267	.292	.316	.341	.365	.390	.414	.439	.463	.487	45°F.
40°F.	.040	.060	.081	.101	.122	.142	.162	.182	.203	.223	.244	.264	.284	.304	.325	.345	.366	.386	.407	40°F.
35°F.	.033	.050	.067	.084	.101	.118	.135	.152	.169	.185	.202	.219	.236	.253	.270	.287	.304	.321	.338	35°F.
32°F.	.030	.045	.060	.075	.090	.105	.120	.135	.150	.165	.181	.196	.211	.226	.241	.251	.271	.286	.301	32°F.
30°F.	.027	.041	.055	.068	.082	.096	.110	.124	.138	.151	.165	.179	.193	.207	.221	.234	.248	.262	.276	30°F.
25°F.	.022	.033	.044	.055	.066	.077	.088	.099	.110	.121	.133	.144	.155	.166	.177	.188	.199	.210	.221	25°F.
20°F.	.017	.026	.035	.043	.052	.061	.070	.079	.088	.096	.105	.114	.123	.132	.141	.149	.158	.176	.176	20°F.
15°F.	.014	.021	.028	.035	.042	.049	.056	.063	.070	.077	.084	.091	.098	.105	.112	.119	.126	.133	.140	15°F.
10°F.	.011	.016	.022	.027	.033	.038	.044	.049	.055	.060	.066	.071	.077	.082	.088	.093	.099	.104	.110	10°F.
+5°F.	.008	.012	.017	.021	.026	.030	.035	.039	.043	.047	.052	.056	.061	.065	.069	.073	.078	.082	.087	+5°F.
0°F.	.006	.009	.013	.016	.020	.023	.027	.030	.034	.037	.041	.044	.048	.051	.055	.058	.061	.064	.068	0°F.
-5°F.	.005	.007	.010	.012	.015	.018	.021	.023	.026	.028	.031	.034	.037	.039	.042	.044	.047	.049	.052	-5°F.
-10°F.	.004	.006	.008	.010	.012	.014	.016	.018	.020	.022	.024	.026	.028	.030	.032	.034	.036	.038	.040	-10°F.
-15°F.	.003	.004	.006	.007	.009	.010	.012	.013	.015	.016	.018	.019	.021	.022	.024	.026	.028	.029	.031	-15°F.

cooling than before, according to whether or not the cooling is carried below the temperature at which the air becomes saturated. Table XXII shows the amount of vapor in pounds per thousand cubic feet of air at different degrees of saturation at different temperatures. At 100° Fahr., for example, 1,000 cubic feet of saturated air will contain 2.82 pounds of water vapor, while at 75° Fahr. the amount is only 1.33 pounds, or less than one-half that quantity, and at 15° Fahr. it is still further reduced to about one-tenth of what it is at 75° Fahr.

In the general case, air cooling involves cooling not only the mechanical mixture of oxygen and nitrogen, but a large quantity of water vapor as well. If the air contains just sufficient moisture so that the cooling brings it to the point of saturation, the heat that must be abstracted from the water vapor will be only that represented by the specific heat of the vapor and the number of degrees cooled through. If it is cooled below the point of saturation, as is usually the case in cold storage practice, not only the specific heat of the vapor but the latent heat of that part of the vapor precipitated as well must be removed. Generally the process is carried still farther and the precipitated moisture is chilled to the freezing point and finally frozen, when not only the specific heat of the liquid but the latent heat of fusion is involved. In case the ice is cooled to a lower temperature the specific heat of the ice is also involved.

EXAMPLE.—It is required to cool 2,000 cu. ft. of air per minute from 80° Fahr. to 36° Fahr. In the following calculations it is assumed that the amount of air to be cooled is 2,000 cu. ft. *before* it is cooled, and not, as it might be construed to mean, 2,000 cu. ft. of cooled air.

For the sake of simplicity the air is assumed to be dry.

Dry air at 80° F. weighs .0731 pounds per cu. ft.

2,000 cu. ft. would weigh	146.2 lbs.
The specific heat of air is	0.2377
B. t. u. required to cool 2,000 cu. ft. 1° F.	34.75
B. t. u. required to cool 2,000 cu. ft. 44° F.	1529.

One ton of refrigeration is sufficient to dispose of heat at the rate of 288,000 B. t. u. per twenty-four hours, or (dividing this number by 1440, the number of minutes in 24 hours) gives the equivalent rate per minute or 200 B. t. u. per minute.

On this basis, the cooling of 2,000 cu. ft. of air per minute from 80° Fahr. to 36° Fahr. would require the expenditure of $1529 \div 200 = 7.64$ tons of refrigeration.

Had the requirements been for 2,000 cu. ft. of *cooled* air, the amount of refrigeration needed would have been 8.36 tons, the difference being accounted for by the difference of weight per cu. ft. of air at 80° Fahr. and 36° Fahr., respectively.

Cold Losses Through Cold Storage Doors

There is no known means of accurately determining loss of refrigeration through the opening of cold storage doors. It is possible that it might be roughly approximated from formulæ giving the flow of gases under slight differences in pressures, in which case some delicate form of draft gauge might be employed to show the excess pressure of the cold air on the inside of the cold storage compartment over that of the outside air. The area through which the outward flow due to the observed difference of pressure would take place would be probably about one-half of that of the opening offered by the door, because in a single opening the upper part would be given up to the inward current of warm air.

While it would be difficult to estimate the velocity at which cold air rushes out of a cold storage compartment, it is apparent that it will increase as the difference between the inside and outside temperatures, and with the increase in height of the cold air column, both of these factors acting to affect an unbalancing of the atmospheric pressures and consequently tending to produce a flow.

In this connection it may be remarked that the circulation of air in cold storage compartments, as well as currents of air entering and leaving the compartment, can be conveniently studied by using smoke as an indicator. It might be possible by means of a puff of smoke and a stop watch, in the absence of a delicate anemometer, to roughly determine the velocity of the air currents. The inward current would have a maximum velocity at the top of the opening and the outward current at the bottom, while somewhere near midway would be found a place with no perceptible current. From this it follows that the volume of air lost through the opening might be determined by multiplying one-half the area of the opening by one-half the maximum velocity. This product of the average velocity in feet per minute and the area of the current will be the number of cubic feet per minute lost through the opening, and since 4,000 cubic feet per minute cooled one degree requires refrigeration at the rate of one ton per

24 hours, it follows that outside air at a temperature of 80° Fahr., rushing into the cold storage compartments to take the place of cold air escaping at a temperature of 40° Fahr., requires an additional ton of refrigeration for every 100 cubic feet of flow. To reduce this excessive loss to a minimum, vestibules sufficiently large to permit one door to be closed before the other is opened are often provided for doors communicating directly with the outside. Where products alone are to be passed, rotary doors, or in the case of ice storage rooms automatically closing swing doors, may be employed to advantage.

Insulation Losses

Good heat insulators are simply poor heat conductors and poor heat insulators, good heat conductors—this property of each being numerically the reciprocal of that of the other. Since all heat insulators are to some extent heat conductors, the flow of heat through insulated walls cannot be prevented but only reduced in proportion to the thickness and efficiency of the insulation employed. The amount of heat that will pass through a square foot of cold storage insulation per 24 hours, like that through other more or less imperfect conductors, is practically proportional to the difference in temperature on the two sides of the insulation and to the efficiency of the material not as a heat insulator but as a heat conductor.

Products cooled to the temperature of the cold storage compartment, where uniform temperatures are maintained, require the expenditure of no further refrigeration.*

The necessity for operating the refrigerating plant for the preservation of such products is therefore due largely to the entrance of heat through the insulated walls of the cold storage compartments and the insulation should be made as efficient as economy will permit. True economy is found at the point where the cost of the refrigeration that would otherwise be lost is balanced up against the cost of the insulation effecting the saving. Obviously, the more it costs to produce a ton of refrigeration the more it is economy to spend for insulation to conserve the refrigeration produced.

* Exceptions to this general rule are products which are fermented while in storage, the process of fermentation giving rise to the evolving of a considerable quantity of heat.

Thermal conductivity varies widely among various so called insulating materials and even with the same material when varying amounts of air and moisture are present. Table XXIII shows the rate of transmission expressed in B. t. u. per square foot per 24 hours per degree difference in temperature between the two sides of the insulation. These values represent efficiencies under best conditions. In making computations for determining the capacity of refrigerating machines it is customary among some builders to increase these values by from 25 to 50 per cent., according to the physical condition of the insulation.

TABLE XXIII.—HEAT CONDUCTIVITIES, *C*, OF COLD STORAGE INSULATION

Transmission in B.t.u. per sq. ft. per degree difference in temperature inside and out per 24 hours. Compiled largely from information published by the Armstrong Cork Co.

Material	B.t.u.
Insulating Slabs.	
1" "Pure Cork Sheets" (Granulated Cork united by heat and pressure)	6.5
1" "Rock Wool Composition Boards" (Waterproofed)	7.4
1" Impregnated Cork Board (Granulated Cork and Asphaltic binder)	8.9
1" Indurated Wood Pulp Board	10.0
Built-up Insulation (Wood and Air space)	
1" American Spruce	16.80
(7/8" Dressed and Matched Spruce) (7/8 Sp.) (paper, 7/8 Sp.) (7/8 Sp. paper, 7/8 Sp.)	4.75
(7/8 Sp. paper, 7/8 Sp.) (1" air space) (7/8 Sp., paper, 7/8 Sp.)	4.25
6 thicknesses 7/8 Sp., 3 papers, 2 air spaces arranged as above	3.45
8 " " 4 " 3 " " " "	2.70
10 " " 5 " 4 " " " "	2.70
(8 thicknesses being 1/2 and 2 thicknesses being 7/8" thick)	
Built-up Insulation Wood, Paper and Fill	
(7/8 Sp. paper, 7/8 Sp.) (7/8 Sp. paper, 7/8 Sp.)	4.75
(" " ") (4" Mineral Wool) (7/8 Sp. paper, 7/8 Sp..)	2.20
(" " ") (8" Mill Shavings, Damp) (7/8 Sp. paper, 7/8 Sp.)	2.10
(" " ") (1" " " Dry) (" " ")	1.35
(" " ") (8" Granulated Cork) " " ")	1.90
(" " ") (1" Pure Sheet Cork) " " ")	3.10
(7/8 Sp. paper) (1" Pure Sheet Cork) (paper, 5/8 Sp.)	3.25
(" ") (2" " ") (" ")	2.60
(" ") (3" " ") (" ")	2.25
(" ") (4" " ") (" ")	1.20
(7/8 Sp.) (1" Pitch) (7/8 Sp.)	4.90
(") (2" ") (")	4.25
Built-up Insulation (*Wood, Paper, Air Space and Fill*)	
(7/8 Sp. paper, 7/8 Sp.) (1" Air Space) (7/8 Sp.) (6" Min. Wool) (7/8 Sp.. paper, 7/8 Sp.)	1.49
(" " ") (" ") (") (6" Gran. Cork) (7/8 Sp., paper, 7/8 Sp.)	1.46
(" " ") (" ") (") (2" Pure Sh. ") (" " ")	1.60
(" " ") (" ") (2" Pure Sheet Cork) (paper, 7/8 Sp.)	2.10
(" " ") (" ") (3" " ") (" ")	1.70
(" " ") (" ") (4" " ") (" ")	1.20
(" " ") (" ") (5" " ") (" ")	.90
Brick Wall and Sheet Cork	
(13" Brick Wall) (2" Pure Sheet Cork)	2.75
(" " ") (4" " ")	1.47

Assuming, for example, a cold storage box $10' \times 10' \times 10'$, the superficial surface exposed is 600 square feet.

The insulation may consist of two courses of $\frac{7}{8}''$ dressed and matched spruce with a course of paper between, a 1" air space and two more courses of spruce with paper between. The conductivity of insulation of this construction is given in the table as 4.25 B. t. u. If the insulation is found to be moist, about 20% may be added to the above value bringing the heat transmission up to about 5 B. t. u.* The 24-hour duty is now found by multiplying 600, the number of square feet surface, by 5, the heat transmission per square foot, giving 3,000 the number of B. t. u. per 24 hours per degree difference in temperature. This multiplied by the difference between the outside and inside temperatures (say 90°–36° or 54°) gives 162,000 B. t. u. as the total heat absorbed.

This divided by 144 B. t. u., the amount of heat required to melt a pound of ice, gives 1,125 pounds or 0.5625 tons as the amount of refrigeration required per 24 hours to make up for insulation losses.

A simple expression for pounds of refrigeration K per 24 hours, per square foot of insulation having a B. t. u. conductivity C per 24 hours per degree difference in temperature $(t-t_1)$, as given in Table XXIV, is

$$K=C\,\frac{(t-t_1)}{144}. \tag{17}$$

Substituting in this expression the values given in the above example gives

$$K=\frac{5\times 54}{144}=1.875$$

which result multiplied by the total square feet of surface, 600, gives 1,125 pounds as before.

Table XXIV gives similar values of K for different insulation conductivities, ranging from 1 to 10 B. t. u. per square foot and for differences in temperature ranging from 40° to 100°.

*As a matter of fact, five B. t. u. per square foot per degree difference in temperature is often employed where the exact value of the insulation cannot be determined, as an approximate factor for estimating the total cold storage duty required for small and medium sized boxes with insulation of the average inferior quality usually employed in market and hotel refrigerators. The amount of refrigerating duty estimated on this basis should be ample to provide not only for the insulation losses but for the cooling of the average small amount of product and the neutralizing of the amount of heat generated by lights, workmen and entering through the opening of doors.

To employ this table in the above example, find constant $K=1.875$ in the horizontal line opposite $(t-t_1)=54°$ and in the vertical column under $C=5$. This factor multiplied by the surface, 600, gives, as before, 1,125 pounds, which divided by 2,000 gives 0.5625 tons of refrigeration as the required capacity to make up for insulation losses.

TABLE XXIV.—Values of Constant, K, Pounds Refrigerating Duty per Square Foot Wall Surface per 24 Hours for Different Insulation Conductivities and Differences in Temperature $(t-t_1)$, Inside and Out.

90° Fahr. (t), assumed outside temperature, minus inside temp. (t_1), $=(t-t_1)$, column 2.

$$K=C\,\frac{(t-t_1)}{144}.$$

Inside Temp. (t_1)	$(t-t_1)$	B.t.u. per Sq. Ft. per Degree Difference in Temperature per 24 Hours									
		1	2	3	4	5	6	7	8	9	10
50	40	.2772	.5564	.8346	1.111	1.391	1.669	1.926	2.222	2.5	2.777
48	42	.2916	.5832	.8749	1.166	1.458	1.75	2.041	2.333	2.635	2.916
46	44	.3055	.6110	.9165	1.222	1.527	1.833	2.139	2.444	2.749	3.055
44	46	.3194	.6388	.9582	1.278	1.597	1.916	2.236	2.555	2.875	3.194
42	48	.3333	.6667	.9999	1.333	1.667	2.	2.333	2.666	3.000	3.333
40	50	.3492	.6944	1.042	1.389	1.736	2.083	2.432	2.777	3.125	3.471
38	52	.3611	.7222	1.083	1.444	1.805	2.167	2.528	2.889	3.249	3.610
36	54	.375	.750	1.125	1.5	1.875	2.25	2.625	3.00	3.375	3.750
34	56	.3899	.7778	1.167	1.556	1.945	2.332	2.729	3.119	3.501	3.883
32	58	.4028	.8056	1.208	1.611	2.014	2.417	2.82	3.222	3.625	4.028
30	60	.4166	.8332	1.25	1.666	2.083	2.5	2.916	3.333	3.749	4.166
28	62	.4306	.8612	1.292	1.722	2.153	2.583	3.014	3.485	3.875	4.306
26	64	.4444	.8888	1.333	1.778	2.222	2.666	3.001	3.555	4.000	4.444
24	66	.4583	.9166	1.375	1.833	2.292	2.75	3.208	3.666	4.125	4.582
22	68	.4722	.9444	1.417	1.889	2.361	2.833	3.305	3.778	4.250	4.722
20	70	.4861	.9722	1.458	1.944	2.431	2.917	3.403	3.889	4.375	4.861
18	72	5.	1.	1.5	2.	2.5	3.	3.5	4.	4.5	5.
16	74	.5139	1.028	1.542	2.056	2.569	3.083	3.597	4.111	4.625	5.139
14	76	.5278	1.056	1.583	2.111	2.639	3.167	3.695	4.222	4.750	5.278
12	78	.5417	1.083	1.625	2.167	2.708	3.25	3.792	4.334	4.875	5.417
10	80	.5556	1.111	1.667	2.222	2.778	3.333	3.889	4.445	5.000	5.555
8	82	.5694	1.139	1.708	2.278	2.847	3.416	3.986	4.555	5.125	5.694
6	84	.5833	1.167	1.75	2.333	2.917	3.5	4.083	4.666	5.250	5.833
4	86	.5972	1.194	1.792	2.389	2.986	3.583	4.180	4.778	5.375	5.972
2	88	.6111	1.222	1.833	2.444	3.056	3.667	4.278	4.889	5.500	6.111
0	90	.625	1.25	1.875	2.5	3.125	3.75	4.375	5.	5.625	6.250
—2	92	.6388	1.277	1.916	2.553	3.193	3 835	4.468	5.110	5.750	6 388
—4	94	.6526	1.301	1.958	2.610	3.262	3.913	4 565	5.22	5.875	6.526
—6	96	.6666	1.333	2.000	2.666	3.333	4 000	4.666	5.333	6.000	6.666
—8	98	.6805	1.361	2.165	2.72	3 470	4.82	4.762	5.442	6.125	6 805
—10	100	.6942	1.388	2.083	2.778	3.192	4.167	4 861	5.555	6.250	6.942

INDEX

T.

Lightning Source UK Ltd.
Milton Keynes UK
UKHW040028060119
335016UK00011B/501/P